DMV Seminar
Band 17

Springer Basel AG

Lennart Ljung
Georg Pflug
Harro Walk

Stochastic Approximation and Optimization of Random Systems

Springer Basel AG

Authors' addresses:

L. Ljung
Linköping University
Department of Electrical Engineering
S–581 83 Linköping
Sweden

G. Pflug
Universität Wien
Institut für Statistik und Informatik
Universitätsstrasse 5
A–1010 Wien
Austria

H. Walk
Universität Stuttgart
Mathematisches Institut A
Pfaffenwaldring 57
D–7000 Stuttgart 80
Germany

Deutsche Bibliothek Cataloging-in-Publication Data

Stochastic approximation and optimization of random systems /
Lennart Ljung; Georg Pflug; Harro Walk. – Basel ; Boston ;
Berlin : Birkhäuser, 1992
  (DMV-Seminar ; Bd. 17)
    ISBN 978-3-7643-2733-0   ISBN 978-3-0348-8609-3 (eBook)
    DOI 10.1007/978-3-0348-8609-3
NE: Ljung, Lennart; Pflug, Georg; Walk, Harra; Deutsche
  Mathematiker-Vereinigung: DMV-Seminar

© 1992 Springer Basel AG
Originally published by Birkhäuser Verlag Basel in 1992
Printed on acid-free paper, directly from the authors' camera-ready manuscripts
ISBN 978-3-7643-2733-0

Preface

The DMV seminar "Stochastische Approximation und Optimierung zufälliger Systeme" was held at Blaubeuren, 28.5.–4.6.1989.

The goal was to give an approach to theory and application of stochastic approximation in view of optimization problems, especially in engineering systems. These notes are based on the seminar lectures. They consist of three parts:

  I. Foundations of stochastic approximation (H. Walk);
 II. Applicational aspects of stochastic approximation (G. Pflug);
III. Applications to adaptation algorithms (L. Ljung).

The prerequisites for reading this book are basic knowledge in probability, mathematical statistics, optimization.

We would like to thank Prof. M. Barner and Prof. G. Fischer for the organization of the seminar. We also thank the participants for their cooperation and our assistants and secretaries for typing the manuscript.

November 1991   L. Ljung, G. Pflug, H. Walk

Table of contents

I   Foundations of stochastic approximation

Harro Walk
University of Stuttgart
Mathematisches Institut A
Pfaffenwaldring 57
D-7000 Stuttgart 80
Federal Republic of Germany

Stochastic approximation or stochastic iteration concerns recursive estimation of quantities in connection with noise contaminated observations. Historical starting points are the papers of Robbins and Monro (1951) and of Kiefer and Wolfowitz (1952) on recursive estimation of zero and extremal points, resp., of regression functions, i.e. of functions whose values can be observed with zero expectation errors.

The Kiefer-Wolfowitz method is a stochastic gradient algorithm and may be described by the following example. A mixture of metals with variable mixing ratio is smelt at a fixed temperature. The hardness of the alloy depends on the mixing ratio characterized by an $x \in \mathbb{R}^k$, but is subject to random fluctuations; let $F(x)$ be the expected hardness. The goal is a recursive estimation of a maximal point of F. One starts with a mixing ratio characterized by X_1 in $\mathbb{R}^k$. Let X_n characterize the mixing ratio in the nth step; for the neighbour points $X_n \pm c_n e_l$ $(l = 1, \ldots, k)$ with $0 < c_n \to 0$ and unit vector e_l (with 1 as l-th coordinate) one obtains the random hardnesses Y'_{nl} and Y''_{nl}, resp. The mixing ratio in the next step is chosen as

$$X_{n+1} := X_n + \frac{c}{n}(\frac{Y'_{nl} - Y''_{nl}}{2c_n})_{l=1,\ldots,k}$$

with $c > 0$. If F is totally differentiable with (Fréchet) derivative DF, the recursion can be formally written as

$$X_{n+1} = X_n + \frac{c}{n}[DF(X_n) - H_n - V_n]$$

with random vectors X_n, H_n, V_n, where $-H_n$ can be considered as a systematic error at using divided differences instead of differential quotients and $-V_n$ as a stochastic error with

$$E(V_n|X_1, H_1, V_1, \ldots, H_{n-1}, V_{n-1}) = 0.$$

Under certain assumptions on F and the errors, (X_n) almost surely (a.s.) converges to a maximal point of F.

For recursive estimation of a minimal point one uses the above formula with $-c/n$ instead of c/n . The recursion is inspired by the gradient method in deterministic optimization. In contrast to most of the deterministic iterative procedures, the correction terms in stochastic approximation procedures which are influenced by random effects, are provided with damping factors c/n or more generally deterministic or random weights (gains) a_n with $0 < a_n \to 0$ (usually) and $\sum a_n = \infty$.

§1   Almost sure convergence of stochastic approximation procedures

Results on almost sure convergence of stochastic approximation processes are often proved by a separation of deterministic (pathwise) and stochastic considerations. The basic idea is to show that a "distance" between estimate and solution itself has the tendency to become smaller. The so-called first Lyapunov method of investigation does not use knowledge of a solution. Thus in deterministic numerical analysis gradient and Newton procedures for minimizing or maximizing F by a recursive sequence (X_n) are investigated by a Taylor expansion of $F(X_{n+1})$ around X_n – a device, which has been used in stochastic approximation for the first time by Blum (1954) and later by Kushner (1972) and Nevel'son and Has'minskii (1973/76, pp. 102 - 106). The second Lyapunov method in its deterministic and stochastic version employs knowledge of a solution. The stochastic consideration often uses "almost supermartingales" (Lemma 1.10).

The following Lemma 1.1 (compare Henze 1966, Pakes 1982) concerns linear recursions and weighted means. Theorems 1.2, 1.7, 1.8 below are purely deterministic and can immediately be formulated in a stochastic a.s. version. In view of applications concerning estimation of functions as elements of a real Hilbert space $\mathbb{H}$ or a real Banach space $\mathbb{B}$ the domain of F is often chosen as $\mathbb{H}$ or $\mathbb{B}$.

1.1. Lemma. *Let be $a_n \in [0,1)$ $(n \in \mathbb{N})$, $\beta_n := [(1-a_n)\dots(1-a_1)]^{-1}$, $\gamma_n := a_n\beta_n$. Then*

a) $\beta_n = 1 + \gamma_1 + \dots + \gamma_n$,

b) $\sum a_n = \infty \Longleftrightarrow \beta_n \uparrow \infty$,

c) for elements X_n, W_n in a linear space the representations

$$X_{n+1} = X_n - a_n X_n + a_n W_n \quad (n \in \mathbb{N}) \quad with \ \ X_1 = 0$$

and

$$X_{n+1} = \beta_n^{-1} \sum_{k=1}^{n} \gamma_k W_k \quad (n \in \mathbb{N})$$

are equivalent.
In the special case $a_n = 1/(n+1)$ one has $\beta_n = n+1$, $\gamma_n = 1$.

1.2. Theorem. *Let $F : \mathbb{H} \to \mathbb{R}$ be bounded from below and have a Fréchet derivative DF. Assume $X_n, W_n, H_n, V_n \in \mathbb{H}$ $(n \in \mathbb{N})$ with*

$$X_{n+1} = X_n - a_n(DF(X_n) - W_n), \ W_n = H_n + V_n,$$

where $a_n \in [0,1)$, $a_n \to 0$ $(n \to \infty)$, $\sum a_n = \infty$. Let β_n, γ_n be defined according to Lemma 1.1.

a) Assume

(1) $DF \in Lip$, i.e. $\underset{K \in \mathbb{R}_+}{\exists} \; \underset{x',x'' \in \mathbb{H}}{\forall} \|DF(x') - DF(x'')\| \leq K\|x' - x''\|$,

(2') $$\sum a_n \|H_n\|^2 < \infty,$$

(2'') $$\beta_n^{-1} \sum_{k=1}^{n} \gamma_k V_k \to 0,$$

(2''') $$\sum a_{n+1} \|\beta_n^{-1} \sum_{k=1}^{n} \gamma_k V_k\|^2 < \infty.$$

Then
$(F(X_n))$ is convergent, $\sum a_n \|DF(X_n)\|^2 < \infty$, $DF(X_n) \to 0$ $(n \to \infty)$.

b) Assume

(3) DF uniformly continuous,

(4) $\underset{\|x\| \to \infty}{\underline{\lim}} \|DF(x)\| > 0$ or $\exists \underset{\|x\| \to \infty}{\lim} F(x) \in \mathbb{R}$,

(5) $\{F(x); x \in \mathbb{H}$ with $DF(x) = 0\}$ nowhere dense in $\mathbb{R}$,

(6) $\underset{R>0}{\forall}$ $\overline{\{F(x); x \in \mathbb{H} \text{ with } \|x\| \leq R \text{ and } \|DF(x)\| \leq \delta\}} \downarrow$

$\{F(x); x \in \mathbb{H}$ with $\|x\| \leq R$ and $DF(x) = 0\}$ $(\delta \to 0)$,

(7) $$\beta_n^{-1} \sum_{k=1}^{n} \gamma_k W_k \to 0.$$

Then
$(F(X_n))$ is convergent, $DF(X_n) \to 0$ $(n \to \infty)$.

PROOF OF THEOREM 1.2a: One uses the first Lyapunov method. Motivated by partial summation, with notations of Lemma 1.1, one has
$$a_n V_n = Z_n + a_n \tilde{V}_n$$
with

$$\tilde{V}_n = \beta_{n-1}^{-1} \sum_{k=1}^{n-1} \gamma_k V_k \to 0 \text{ according to (2'')},$$

$$Z_n = \tilde{V}_{n+1} - \tilde{V}_n.$$

Thus

$$X_{n+1} = X_n - a_n DF(X_n) + Z_n + a_n \tilde{V}_n + a_n H_n.$$

Setting

$$X_1' := X_1,$$
$$X_{n+1}' := X_n' - a_n DF(X_n) + a_n \tilde{V}_n + a_n H_n,$$

one obtains

$$X_{n+1}' - X_{n+1} = -(Z_1 + \ldots + Z_n) = -\tilde{V}_{n+1} \to 0,$$
$$X_{n+1}' = X_n' - a_n DF(X_n') + a_n h_n$$

with

$$h_n = DF(X_n') - DF(X_n) + \tilde{V}_n + H_n,$$
$$\sum a_n \|h_n\|^2 < \infty \text{ because of (1), (2'''), (2').}$$

Therefore and because of the Lipschitz condition on DF and $(2''')$ it suffices to prove the assertions for the special case $V_n = 0$.

From the recursion formula one obtains

$$F(X_{n+1}) - F(X_n)$$
$$= -a_n \|DF(X_n)\|^2 + a_n(DF(X_n), H_n)$$
$$+ a_n \int_0^1 (DF(X_n + t\, a_n[-DF(X_n) + H_n]) - DF(X_n), -DF(X_n) + H_n)dt$$

and, by the Lipschitz condition on F,

$$F(X_{n+1}) \le F(X_n) - \tfrac{1}{4}a_n \|DF(X_n)\|^2 + a_n \|H_n\|^2$$

for sufficiently large n. Now $(2')$ yields convergence of $(F(X_n))$. Further, because F is bounded from below, one obtains convergence of $\sum a_n \|DF(X_n)\|^2$. For $DF(X_n) \to 0$ an indirect proof is given. Assume existence of an $\varepsilon > 0$ with $\|DF(X_n)\| \ge \varepsilon$ for infinitely many n. Then, because of $(2')$ and the foregoing result, an N exists with $\|DF(X_N)\| \ge \varepsilon$,

$$\sum_{k=N}^{\infty} a_k \|DF(X_k)\|^2 \le \tfrac{\varepsilon^2}{12K}, \quad \sum_{k=N}^{\infty} a_k \|H_k\|^2 \le \tfrac{\varepsilon^2}{12K},$$

By induction, $\|DF(X_k)\| \ge \varepsilon/2$ for all $k \ge N$ will be shown, which is, because of $\sum a_n = \infty$, in contrast to the foregoing result. Let $\|DF(X_k)\| \ge \varepsilon/2$ for $k = N, \ldots, n$. The recursion formula yields

$$X_{n+1} = X_N - \sum_{k=N}^{n} a_k DF(X_k) + \sum_{k=N}^{n} a_k H_k,$$

thus

$$\|X_{n+1} - X_N\| \le \sum_{k=N}^{n} a_k \|DF(X_k)\| + \sum_{k=N}^{n} a_k \|H_k\|\chi_{[\|H_k\| \le \varepsilon/2]}$$
$$+ \sum_{k=N}^{n} a_k \|H_k\|\chi_{[\|H_k\| > \varepsilon/2]}$$

$$\leq \quad \frac{4}{\varepsilon}\sum_{k=N}^{n} a_k \|DF(X_k)\|^2 + \frac{2}{\varepsilon}\sum_{k=N}^{n} a_k \|H_k\|^2$$

$$\leq \quad \frac{\varepsilon}{2K},$$

$$\|DF(X_{n+1})\| \quad \geq \quad \|DF(X_N)\| - \|DF(X_{n+1}) - DF(X_N)\|$$

$$\geq \quad \varepsilon - K\|X_{n+1} - X_N\|$$

$$\geq \quad \frac{\varepsilon}{2}.$$

$\square$

SKETCH OF THE PROOF OF THEOREM 1.2b: Similarly to the first part of the proof of Theorem 1.2a one shows that without loss of generality $V_n = 0$, $H_n \to 0$ can be assumed. Let the indices be sufficiently large. The assumptions yield that as long as the coercitive part $DF(X_n)$ is small, the values of $F(X_n)$ lie in a small interval; on the other side, as long as $\|DF(X_n)\|$ is bounded away from zero, $F(X_n)$ is decreasing. Thus convergence of $(F(X_n))$ is obtained. – The indirect proof of $DF(X_n) \to 0$ is similar to the corresponding argument for Theorem 1.2a. For N sufficiently large and

$$\|DF(X_N)\| \geq \varepsilon > 0, \quad \|DF(X_k)\| \geq \tfrac{\varepsilon}{2} \quad (k = N+1, \ldots, n)$$

one would obtain

$$\|X_{n+1} - X_N\| \leq \tfrac{3}{\varepsilon}\sum_{k=N}^{n} a_k \|DF(X_k)\|^2 \leq \tfrac{9}{\varepsilon}[-F(X_{n+1}) + F(X_N)]$$

and

$$\|DF(X_{n+1}) - DF(X_N)\| < \tfrac{\varepsilon}{2}$$

by use of the foregoing result, $H_k \to 0$ $(k \to \infty)$ and uniform continuity of DF, thus

$$\|DF(X_k)\| \geq \tfrac{\varepsilon}{2} \text{ for all } k \geq N,$$

which is, because of $\sum a_k = \infty$, in contrast to the assumption that F is bounded from below. $\square$

1.3. Remark. a) For stochastic versions of Theorem 1.2a in $\mathbb{R}^k$ with martingales compare the literature mentioned before. Theorem 1.2b is a variant of a result of Ljung (1978) concerning $\mathbb{R}^k$ and will be applied in §2 and §4.
b) Let $\mathbb{H}$ be separable. Assumption $(2'')$ in Theorem 1.2a is in the case $a_n = 1/(n+1)$ an ergodicity assumption. For general a_n it is implied by convergence of $\sum a_n V_n$ according to the Kronecker lemma. Sufficient conditions in the case of square integrability of the random elements V_n are

1) (V_n) martingale difference sequence with $\sum a_n^2 E\|V_n\|^2 < \infty$,
2) V_n uncorrelated with $EV_n = 0$, $\sum a_n^2 E\|V_n\|^2 (\log n)^2 < \infty$ (Rademacher-Menshov, see Révész 1968),
3) $EV_n = 0$ and $|EV_n V_m| \leq$ const $(\beta_n^p + \beta_m^p)/(1 + |\beta_n - \beta_m|^q)$ with $0 \leq 2p < q < 1$ (compare Cramér and Leadbetter 1967 pp. 94 - 96). See also Remark 1.13b. For (7) in Theorem 1.2b analogous conditions hold.

c) In Theorem 1.2b assumption (4) can be replaced by the assumption that (X_n) is bounded. Here uniform continuity of DF can be replaced by uniform continuity of DF on bounded sets, in the case $\mathbb{H} = \mathbb{R}^k$ by continuous differentiability of F. Further in the case $\mathbb{H} = \mathbb{R}^k$ apparently (6) can be dropped and (5) is fulfilled under the other assumptions, if F is k-times continuously differentiable, according to the Morse-Sard theorem (see e.g. Milnor 1965/1972).

1.4. Remark. a) Assume for Theorem 1.2a that in the n-th iteration step only $\phi_n DF(X_n)$ can be observed with error $-\phi_n W_n$, where the operators $\phi_n : \mathbb{H} \to \mathbb{H}$ yield projections onto finite-dimensional subspaces $H^n \uparrow$ (Nixdorf 1984; compare also Goldstein 1988 with references and applications). Then for the corresponding modification of the recursion with assumptions $(2')$, $(2'')$, $(2''')$ for $\phi_n W_n$, $\phi_n H_n$, ϕV_n instead of W_n, H_n, V_n, resp., one obtains convergence of $(F(X_n))$, further

$$\sum a_n \|\phi_n DF(X_n)\|^2 < \infty$$

and, in the spirit of the Ritz-Galerkin method,

$$\phi_n DF(X_n) \to 0 \quad (n \to \infty).$$

b) Theorem 1.2b also holds for a recursion with $g : \mathbb{H} \to \mathbb{H}$ instead of DF under the following assumption:
$DF(x) = 0 \iff g(x) = 0 \iff (DF(x), g(x)) = 0$,
$(DF(x), g(x)) \geq 0 \quad (x \in \mathbb{H})$,
boundedness of the quotients of $\|DF\|$, $\|g\|$, $\sqrt{(DF, g)}$ as far as defined.

The following theorem concerns a Kiefer-Wolfowitz process (X_n) for minimizing a regression function $F : \mathbb{R}^k \to \mathbb{R}$. Its proof is a stochastic analogue of the proof of Theorem 1.2a and uses that $(F(X_n) - \inf\{F(x); x \in \mathbb{R}^k\})$ is a nonnegative almost supermartingale (see Lemma 1.10, compare the proof of Theorem 1.9 below). In the special case $a_n = c/n$ with $c > 0$, without loss of generality $c = 1$, and $c_n = d n^{-\gamma}$ with $d > 0, 0 < \gamma < 1/2$ the theorem immediately follows from Theorem 1.2a together with Remark 1.3b by noticing $\|H_n\| \leq$ const c_n for the systematic errors $-H_n$.

1.5. Theorem. *Let $F : \mathbb{R}^k \to \mathbb{R}$ be measurable and bounded from below. Assume existence of $DF \in Lip$. Let X_n, $V_n' = (V_{n1}', \ldots, V_{nk}')$, $V_n'' = (V_{n1}'', \ldots, V_{nk}'')$ ($n \in \mathbb{N}$) be k-dimensional random vectors with*

$$X_{n+1} = X_n - a_n \tilde{D}_n F(X_n), \ n \in \mathbb{N},$$

where

$$0 \leq a_n \to 0 \ (n \to \infty), \ \sum a_n = \infty,$$

$$(\tilde{D}_n F(X_n))_l := \frac{F(X_n + c_n e_l) - V_{nl}' - [F(X_n - c_n e_l) - V_{nl}'']}{2c_n},$$

e_l l-th unit vector in $\mathbb{R}^k$ ($l = 1, \ldots, k$),

$$c_n > 0, \ \sum a_n c_n^2 < \infty, \ \sum a_n^2 c_n^{-2} < \infty,$$

$$\sup_n E(\|V_n'\|^2 + \|V_n''\|^2) < \infty,$$

$$E(V_n'|\mathcal{F}_n) = E(V_n''|\mathcal{F}_n) = 0, \ \mathcal{F}_n = \mathcal{F}(X_1, V_1', V_1'', \ldots, V_{n-1}', V_{n-1}'').$$

Then a.s. the assertions of Theorem 1.2a hold.

1.6. Remark. If in Theorem 1.2 with $\mathbb{H} = \mathbb{R}^k$ or in Theorem 1.5 $F : \mathbb{R}^k \to \mathbb{R}$ is assumed to be inf-compact, i.e. $\{x \in \mathbb{R}^k; \; F(x) \leq \lambda\}$ compact, then (a.s.):
each accumulation point of (X_n) is stationary point of F,
dist$(X_n$, set of stationary points of $F) \to 0$,
the set of accumulation points of (X_n) is connected.

    Theorem 1.7 (Clark 1984) and Theorem 1.8 (Schwabe 1986), both of deterministic nature, are applicable especially to the Robbins-Monro procedure for recursive estimation of a zero point of a regression function f ($f_n = f$ for all n in Theorem 1.8) by a stochastic sequence (X_n), the Robbins-Monro process, and to the Kiefer-Wolfowitz procedure. As to the latter procedure for recursive estimation of a stationary point of a differentiable regression function F, Theorem 1.7 uses $f = DF$ and comprehends systematic and stochastic errors as $-W_n$, and Theorem 1.8 uses

$$f_n(x) = \frac{F(x + c_n) - F(x - c_n)}{2c_n} \; , \; x \in \mathbb{R}, \; n \in \mathbb{N},$$

with span $2c_n$, and stochastic errors $-V_n$. Theorem 1.8 more generally concerns a so-called dynamic stochastic approximation procedure in $\mathbb{R}$ (compare also Dupač 1966 as an early reference). Both theorems stress the narrow connection between strong consistency of the estimates and the convergence behaviour of the errors under typical regularity conditions on f (especially f and f_n, resp., quasilinear and bounded away from zero on certain subdomains).

1.7. Theorem. *Assume $f : \mathbb{R}^k \to \mathbb{R}^k$, $\vartheta \in \mathbb{R}^k$ with*

    $f \in Lip,$

$$\underset{c > 0}{\exists} \;\; \underset{x \in \mathbb{R}^k}{\forall} \;\; (f(x), \, x - \vartheta) \geq c\|x - \vartheta\|^2.$$

Let $X_n, W_n \in \mathbb{R}^k$ ($n \in \mathbb{N}$) with

    $X_{n+1} = X_n - a_n(f(X_n) - W_n),$

where

    $a_n = \rho_n/n$ *with* $0 < \underline{\lim} \, \rho_n \leq \overline{\lim} \, \rho_n < \infty.$

Then

$$X_n \to \vartheta \iff \frac{1}{n} \sum_{k=1}^{n} W_k \to 0.$$

    The proof uses the second Lyapunov method and an embedding into a differential equation of the form

$$\dot{X}(t) = -a(t)f(X(t)) + a(t)\tilde{W}(t).$$

1.8. Theorem. *Assume $f_n : \mathbb{R} \to \mathbb{R}$, $\vartheta \in \mathbb{R}$ with*

$$\underset{c \in \mathbb{R}_+}{\exists} \;\; \underset{N}{\exists} \;\; \underset{n \geq N}{\forall} \;\; \underset{x \in \mathbb{R}}{\forall} \;\; |f_n(x)| \leq c(1 + |x|) \; (quasilinearity),$$

$$\underset{0 < \delta_1 \leq \delta_2}{\forall} \;\; \underset{n}{\underline{\lim}} \;\; \underset{\delta_1 \leq |x - \vartheta| \leq \delta_2}{\inf} \;\; |f_n(x)| > 0,$$

$$\underset{\varepsilon>0}{\forall}\ \underset{N}{\exists}\ \underset{n\geq N}{\forall}\ \underset{x\in\mathbb{R}}{\forall}\quad |x-\vartheta|\geq \varepsilon \Longrightarrow f_n(x)(x-\vartheta)>0.$$

Let $X_n, V_n \in \mathbb{R}\ (n \in \mathbb{N})$ with

$$X_{n+1} = X_n - a_n(f_n(X_n) - V_n),$$

where

$$a_n \geq 0,\ a_n \to 0\ (n \to \infty),\ \sum a_n = \infty.$$

Then

$$\sum a_n V_n\ convergent \Longrightarrow X_n \to \vartheta.$$

The proof uses the second Lyapunov method and a contraction (as to the latter compare also Dvoretzky 1956).

The following theorem of stochastic nature is essentially due to Gladyshev (1965) and yields a relatively simple martingale access to Robbins-Monro and Kiefer-Wolfowitz processes.

1.9. Theorem. *Let the real separable Hilbert space* $\mathbb{H}$ *be endowed with the Borel* σ-*algebra and* $f : \mathbb{H} \to \mathbb{H}$ *be measurable. Assume* $\vartheta \in \mathbb{H}$ *(usually, but not necessarily with* $f(\vartheta) = 0$). *Let further* X_n, W_n, H_n, V_n *be* $\mathbb{H}$-*valued random variables* $(n \in \mathbb{N})$ *with*

$$X_{n+1} = X_n - a_n(f(X_n) - W_n),\ W_n = H_n + V_n,$$

where

$$a_n \geq 0,\ \sum a_n^2 < \infty,\ \sum a_n = \infty.$$

Assume

(7) $\quad \underset{c\in\mathbb{R}_+}{\exists}\ \underset{x\in\mathbb{H}}{\forall}\quad \|f(x)\| \leq c(1 + \|x\|),$

(8) $\quad \underset{K\in[1,\infty)}{\forall}\quad \inf\{(f(x), x - \vartheta);\ x \in \mathbb{H}\ with\ \frac{1}{K} \leq \|x - \vartheta\| \leq K\} > 0,$

(9) $\quad \underset{n}{\forall}\quad H_n, V_n\ square\ integrable\ with$

$$\sum a_n E\|H_n\| < \infty,\ \sum a_n^2 E\|H_n\|^2 < \infty,$$
$$E(V_n|X_1, H_1, V_1, \dots, H_{n-1}, V_{n-1}) = 0,\ \sum a_n^2 E\|V_n\|^2 < \infty.$$

Then

$$X_n \to \vartheta\ (n \to \infty)\ a.s.$$

Theorem 1.9 concerns in the special case $H_n = 0\ (n \in \mathbb{N})$ the usual Robbins-Monro process, where $-V_n$ can be interpreted as a random error in the observation of $f(X_n)$. Especially for $f = id$, $a_n = 1/n$ one has

$$X_{n+1} = \frac{1}{n}\sum_{j=1}^{n} V_j \quad (\to 0\ a.s.\ ).$$

In the case $\mathbb{H} = \mathbb{R}$, for continuous f, (8) is equivalent to

$$f(x) \overset{\geq}{<} 0\ for\ x \overset{\geq}{<} \vartheta.$$

The proof of Theorem 1.9 given below uses the second Lyapunov method in connection with the sequence $(\|X_{n+1} - \vartheta\|^2)$ instead of the sequence $(F(X_{n+1}))$ in the proof of Theorem 1.2. The sequence is considered as a nonnegative almost supermartingale in the sense of Lemma 1.10 due to Robbins and Siegmund (1971). These authors use the argument also for more complicated situations. Further applications concern e.g. optimization under constraints (Kushner and Sanvicente 1974, Hiriart-Urruty 1977; see §3) and cluster analysis (Pflug 1980).

1.10. Lemma. *Let (U_n) be a sequence of integrable nonnegative real random variables on a probability space $(\Omega, \mathcal{A}, P)$ and $(\mathcal{A}_n)$ be a nondecreasing sequence of sub-σ-algebras of $\mathcal{A}$ with $\mathcal{A}_n - \mathcal{B}_+$-measurability of U_n $(n \in \mathbb{N})$. Assume*

$$E(U_{n+1}|\mathcal{A}_n) \leq (1 + \alpha_n)U_n + \beta_n \quad (n \in \mathbb{N}),$$

where α_n, β_n are integrable nonnegative real random variables on $(\Omega, \mathcal{A}, P)$ with $\mathcal{A}_n - \mathcal{B}_+$-measurability and $\sum E\alpha_n < \infty$, $\sum E\beta_n < \infty$. Then the sequence (U_n), a so-called nonnegative almost supermartingale, converges a.s.

PROOF OF LEMMA 1.10: One sets

$$Z_n := \left[\prod_{i=1}^{n-1}(1 + \alpha_i)\right]^{-1} U_n - \sum_{k=1}^{n-1}\left[\prod_{i=1}^{k}(1 + \alpha_i)\right]^{-1} \beta_k$$

and obtains that $(-Z_n)$ is a submartingale. Further boundedness of $(E|Z_n|)$ is shown by considering Z_n^- and Z_n^+. The submartingale convergence theorem yields a.s. convergence of (Z_n), from which the assertion follows. □

PROOF OF THEOREM 1.9: Without loss of generality one assumes $\vartheta = 0$, boundedness of X_1 and thus square integrability of X_n and $f(X_n)$. From the recursion and (7) one obtains

(10) $\|X_{n+1}\|^2 \leq (1 + 4c^2a_n^2 + a_n\|H_n\|)\|X_n\|^2$
$\qquad + 4a_n^2(c^2 + \|H_n\|^2 + \|V_n\|^2) + a_n\|H_n\|$
$\qquad - 2a_n(f(X_n), X_n) + 2a_n(X_n, V_n).$

For $U_n := \|X_n\|^2$,

$\mathcal{A}_n \quad := \quad \mathcal{F}(X_1, H_1, V_1, \ldots, H_{n-1}, V_{n-1})$
$\qquad = \quad \mathcal{F}(X_1, \ldots, X_n, H_1, \ldots, H_{n-1}, V_1, \ldots, V_{n-1}),$
$\alpha_n \quad := \quad 4c^2a_n^2 + a_n E(\|H_n\| | \mathcal{A}_n),$
$\beta_n \quad := \quad 4a_n^2[c^2 + E(\|H_n\|^2 | \mathcal{A}_n) + E(\|V_n\|^2 | \mathcal{A})] + a_n E(\|H_n\| | \mathcal{A}_n)$

the general assumptions of Lemma 1.10 are fulfilled because of (8), (9), (10). Thus a nonnegative real random variable T exists with

(11) $\|X_n\|^2 \to T$ a.s.

It remains to show $T = 0$ a.s. Using (9) and a.s. boundedness of $(\|X_n\|)$, one obtains a.s. convergence of $\sum a_n(X_n, V_n)$ by truncation and then

(12) $\sum a_n(f(X_n), X_n) < \infty$ a.s.

by (10) and (9). Now a pathwise consideration shows

$P[T > 0] \leq P[\underline{\lim}(f(X_n), X_n) > 0] \leq P[\sum a_n(f(X_n), X_n) = \infty]$

because of (11), (8) and $\sum a_n = \infty$, which together with (12) yields the assertion.   □

Now an application of Theorems 1.5 and 1.9 to minimization of a convex regression function is given.

1.11. Theorem. *Let the assumptions of Theorem 1.5 be fulfilled.*

a) *Assume F strictly convex with minimal point $\vartheta \in \mathbb{R}^k$ or only, if $c_n \to 0$ and $\sum a_n c_n < \infty$,*

$$\underset{\vartheta \in \mathbb{R}^k}{\exists} \quad \underset{x \in \mathbb{R}^k \setminus \{\vartheta\}}{\forall} \quad (Df(x),\, x - \vartheta) > 0.$$

Then
$$X_n \to \vartheta \ a.s.$$

b) *Assume F inf-compact and convex (with not necessarily unique minimal point), $c_n \to 0$, $\sum a_n c_n < \infty$. Then (X_n) a.s. converges to a (generally random) minimal point of F.*

PROOF: a) As to the first part, by Theorem 1.5 one obtains $DF(X_n) \to 0$, convergence of $(F(X_n))$ a.s. and then indirectly proves a.s. boundedness of (X_n) because of the additional assumption. This together with $DF(X_n) \to 0$ a.s. yields $X_n \to \vartheta$ a.s. The second part immediately follows from Theorem 1.9 with $f = DF$ by noticing $\|H_n\| \leq$ const c_n for the systematic errors $-H_n$.
b) The idea is to show that a.s. the sequence (X_n) is seized by an accumulation point (compare Pflug 1981). As before one uses a representation

$$X_{n+1} = X_n - a_n\, DF(X_n) + a_n H_n + a_n W_n$$

with systematic errors $-H_n$ and stochastic errors $-W_n$. By Theorem 1.5 one obtains $DF(X_n) \to 0$ and convergence of $(F(X_n))$ and thus boundedness of (X_n) a.s., because F is inf-compact. Further $\sum a_n^2 \|W_n\|^2 < \infty$, $\sum a_n W_n$ and $\sum a_n(X_n, W_n)$ a.s. converge (compare the analogous argument in the proof of Theorem 1.9). The following pathwise consideration neglects a set of probability measure zero. Let X^* be an accumulation point of (X_n). Then $DF(X^*) = 0$; because F is convex, X^* is a minimal point of F and $(X_n - X^*,\, DF(X_n)) \geq 0$. The above recursion yields

$$\|X_{n+1} - X^*\|^2 \leq \|X_n - X^*\|^2 + D_n$$

with convergence of $\sum D_n$. Thus $(\|X_n - X^*\|)$ is convergent, of course to 0, which yields the assertion.   □

Problems of a.s. convergence of stochastic approximation processes were treated via ordinary differential equations and their stability theory by Ljung (1977), Kushner and Clark (1978), Métivier and Priouret (1984, 1987). This approach denoted as the O. D. E. method shall be described in the following.

Assume that $\vartheta \in \mathbb{R}^k$ is a zero-point of the continuous function $h : \mathbb{R}^k \to \mathbb{R}^k$. Let $x(\cdot, \eta)$, $\eta \in \mathbb{R}^k$, be the solution of the differential equation

$$\dot{x}(t) = -h(x(t)), \quad t \in \mathbb{R},$$

with $x(0, \eta) = \eta$. The equilibrium $x(t) \equiv \vartheta$ is called stable, if for each $\varepsilon > 0$ a $\delta > 0$ exists such that $\|\eta - \vartheta\| < \delta$ implies $\|x(t, \eta) - \delta\| < \varepsilon$ for all $t \geq 0$. It is called (locally) asymptotically stable, if $x(t, \eta) \to 0$ $(t \to \infty)$ for each element η of some neighbourhood of ϑ. The set $A(\vartheta)$ of all such η is denoted as the domain of attraction of ϑ. In the case $A(\vartheta) = \mathbb{R}^k$, ϑ is called globally asymptotically stable. The equilibrium ϑ is asymptotically stable [globally asymptotically stable], if a totally differentiable function V, so-called Lyapunov function, exists on a ball around ϑ with
$V(\vartheta) = 0$, $V(x) > 0$ for $x \neq \vartheta$ (describing a "distance" between x and ϑ),
$(-h(x), DV(x)) < 0$ for $x \neq \vartheta$
[and even
$$V(x) \geq \phi(\|x\|),$$
where $\phi : \mathbb{R}_+ \to \mathbb{R}_+$ is continuous, nondecreasing with $\phi(r) \to \infty$ $(r \to \infty)$].

The O. D. E. method is based on the following deterministic lemma, which yields a connection between the asymptotic behaviour of a recursively defined sequence and the stability behaviour of a corresponding differential equation.

1.12. Lemma. *Let* $h : \mathbb{R}^k \to \mathbb{R}^k$ *be continuous,* $a_n \in \mathbb{R}_+$ *with* $a_n \downarrow 0$, $\sum a_n = \infty$, *and* $X_n, U_n \in \mathbb{R}^k$ *with*
$$X_{n+1} = X_n - a_n h(X_n) + a_n U_n.$$
Assume

(X_n) *bounded,*

$$\sup\{|\sum_{i=n}^{m} a_i U_i|; \ m \in \{n, \ldots, \max\{j; \ a_{n-1} + \ldots + a_{j-1} \leq T\}\}\}$$
$\to 0$ $(n \to \infty)$ *for each* $T > 0$.

Further assume that ϑ *is a locally asymptotically stable point of the differential equation* $\dot{X} = -h(X)$ *in* $\mathbb{R}^k$ *with domain of attraction* $A(\vartheta)$. *If a compact set* $K \subset A(\vartheta)$ *exists with* $X_n \in K$ *for infinitely many* n, *then*
$$X_n \to \vartheta \ (n \to \infty).$$

1.13. Remark. a) By Lemma 1.1 with its notations and partial summation, one gets that the assumption on (U_n) in Lemma 1.12 is fulfilled, if
$$\tfrac{1}{\beta_n} \sum_{i=1}^{n} \gamma_i U_i \to 0 \ (n \to \infty).$$

b) In applications of the deterministic Lemma 1.12 on stochastic approximation with k-dimensional random vectors X_n and U_n, one usually separates a systematic error $-H_n$ with $H_n \to 0$ (a.s.) from $-U_n$. Then one tries to verify the assumption on U_n, now formulated for $E_n := U_n - H_n$, by use of martingale inequalitites or by showing that (E_n) is a so-called convergence system (Chen, Lai and Wei 1981; Lai 1985), i.e. $\sum a_n E_n$ is a.s. and L_2-convergent for each sequence (a_n) in $\mathbb{R}$ with $\sum a_n^2 < \infty$. Examples for convergence systems are L_2-bounded martingale difference sequences, weakly multiplicative sequences, stationary Gaussian sequences with bounded spectral densities.

SKETCH OF THE PROOF OF LEMMA 1.12: (Besides the above mentioned references to the O. D. E. method compare Lai 1985.) Setting $t_n := \sum_{i=1}^{n} a_i$ corresponds to change of variables in differential equations. Further one defines $X(t) := X_{n+1}$ for $t = t_n$ and then by linear interpolation for $t \in (t_n, t_{n+1})$, where $X(0) := X_1$, $t_0 := 0$. If the error terms $-U_n$ would be neglected, then

$$\frac{X(t_n) - X(t_{n-1})}{t_n - t_{n-1}} = -h(X(t_{n-1}))$$

would hold – an approximation of the differential equation $\dot{X} = -h(X)$. Now one defines $U(0) := 0$, $U(t) := \sum_{i=1}^{n} a_i U_i$ for $t = t_n$ and then by linear interpolation for (t_n, t_{n+1}). With $X^*(t) = X_{n+1}$ for $t_n \leq t < t_{n+1}$ the recursion can be written in the form

$$X(t) = X(0) - \int_0^t h(X^*(s))ds + U(t), \ t \geq 0.$$

With

$$\overline{X}_n(t) := X(t + t_n) \text{ for } t \geq -t_n, \ \overline{X}_n(t) := X_1 \text{ for } t \leq -t_n,$$

$$U_n(t) := U(t + t_n) - U(t_n) \text{ for } t \geq -t_n, \ U_n(t) := -U(t_n) \text{ for } t \leq -t_n,$$

$$D_n(t) := \int_0^t [h(X^*(t_n + s)) - h(h(t_n + s))]ds \text{ for } t \geq t_n$$

one obtains

$$\overline{X}_n(t) = X_n(0) - \int_0^t h(\overline{X}_n(s))ds + U_n(t) - D_n(t), \quad t \geq t_n.$$

The sequence of functions $\overline{X}_n : \mathbb{R}_+ \to \mathbb{R}^k$ is relatively compact with respect to the topology of uniform convergence on bounded intervals, and the limit of each convergent subsequence solves the differential equation $\dot{X} = -h(X)$. This leads to the assertion. $\square$

Métivier and Priouret (1984) apply Lemma 1.12 to the following stochastic iteration. Let X_n, Y_n ($n \in \mathbb{N}$) be k-dimensional random vectors with

$$X_{n+1} = X_n - a_n f(X_n, Y_{n+1}),$$

where $f : \mathbb{R}^k \times \mathbb{R}^k \to \mathbb{R}^k$ is measurable and a_n chosen as above. Let be given X_1, Y_1 and the transition probability measures

$$\prod_x(\cdot, \cdot) : (\mathbb{R}^k, \mathcal{B}_k) \to \mathbb{R}, \ x \in \mathbb{R}^k,$$

with

$$P[Y_{n+1} \in B | X_1, \ldots, X_n, Y_1, \ldots, Y_n] = \prod_{X_n}(Y_n, B), \ B \in \mathcal{B}_k, \ n \in \mathbb{N}.$$

Thus (Y_n) can be considered as a Markov chain controlled by (X_n). It is assumed that for each $x \in \mathbb{R}^k$ the Markov chain with transition probabilty measure $\prod_x$ possesses a unique invariant probability measure Γ_x. Further the authors assume that certain regularity conditions of f, $\prod_x$ and the solution v_x of the Poisson equation

$$v_x(\cdot) - \int\limits_{\mathbb{R}^k} v_x(y) \textstyle\prod_x(\cdot, dy) = f(x, \cdot) - \int\limits_{\mathbb{R}^k} f(\cdot, y)\Gamma_x(dy)$$

are fulfilled. They write the above recursion in the form

$$X_{n+1} = X_n - a_n h(X_n) + a_n U_n$$

with

$h(x) = \int f(x, y)\Gamma_x(dy)$ (an averaging principle),
$U_n = -[f(X_n, Y_{n+1}) - h(X_n)]$

and verify the assumptions of Lemma 1.12. – The authors then give an application especially to an algorithm with linear dynamics (Ljung 1977) in the form

$$X_{n+1} = X_n - a_n f(X_n, Y_{n+1}),$$
$$Y_{n+1} = A(X_n)Y_n + B(X_n)W_{n+1},$$

where $A(x)$, $B(x)$ constitute two families of $k \times k$-matrices, X_n, Y_n, W_n are k-dimensional random vectors with independant identically distributed W_n. In this case under regularity conditions implying a.s. convergence of the series below, they obtain that $\prod_x(z, \cdot)$ is the distribution of $A(x)z + B(x)W_1$ ($z \in \mathbb{R}^k$) and Γ_x is the distribution of

$$\sum_{k=0}^{\infty} A^k(x)B(x)W_{k+1}.$$

Further the authors give an application to algorithms driven by an ergodic process. They also treat the problem of a.s. convergence to a unique solution ϑ of $\dot{x}(t) = -h(x(t))$ for linear h under ergodicity assumptions (compare §2 below) and in a later paper (1987) in a general context without boundedness assumptions.

For algorithms driven by an ergodic process a systematic use of the averaging method in the theory of differential equations (see Sanders and Verhulst 1985, §§3, 4) allows to establish the following result, in a deterministic formulation, where boundedness assumptions are avoided.

1.14. Theorem. *Let $f(n, \cdot) : \mathbb{B} \to \mathbb{B}$ possess a Fréchet derivative $Df(n, \cdot)$, $n \in \mathbb{N}$. Assume that a sequence (d_n) in $\mathbb{R}_+$ exists with*

$$\sup_z \|Df(n, z)\| \le d_n, \; n \in \mathbb{N},$$

and

convergence of $(\frac{1}{n}\sum_{k=1}^{n} d_k)$,

and that a Fréchet differentiable function $h : \mathbb{B} \to \mathbb{B}$ and a zero sequence (g_n) in $\mathbb{R}_+$ exist with

$$\|\tfrac{1}{n}\sum_{k=1}^{n} f(k, z) - h(z)\| \le g_n \cdot (1 + \|z\|), \; n \in \mathbb{N}, \; z \in \mathbb{B}.$$

Assume further that an operator $A \in L(\mathbb{B})$ and a $\delta > 0$ exist such that

$$\delta < \min\{re\,\lambda; \; \lambda \in spec(A)\},$$

thus

$$\|e^{-At}\| \le N_\delta \, e^{-\delta t}, \ t \in \mathbb{R}_+,$$

for some $N_\delta \ge 1$, and that

$$\sup_z \|Dh(z) - A\| \le \delta/N_\delta$$

(trivially for $A = \delta^* I$ with $\delta^* > \delta$ and also for $\mathbb{B} = \mathbb{H}$ and A symmetric, $N_\delta = 1$ can be chosen). Let (X_n) be a sequence in $\mathbb{B}$ with

$$X_{n+1} = X_n - \tfrac{1}{n} f(n, X_n), \ n \in \mathbb{N}.$$

Then

$$X_n \to \vartheta \ (n \to \infty),$$

where ϑ is the unique zero point of h.

1.15. Remark. In the case

$$f(n, z) = b_n + \sum_{j=1}^{M} A_{nj} \psi_j(z), \ n \in \mathbb{N}, \ z \in \mathbb{B},$$

with $b_n \in \mathbb{B}$, $A_{nj} \in L(\mathbb{B})$, Fréchet differentiable $\psi_j : \mathbb{B} \to \mathbb{B}$ ($j = 1, \dots, M$; $M \in \mathbb{N}$) the ergodicity assumptions in Theorem 1.14 are fulfilled, if $b \in \mathbb{B}$, $A_j \in L(\mathbb{B})$ exist with

$$\|\tfrac{1}{n} \sum_{k=1}^{n} b_k - b\| \to 0, \ \|\tfrac{1}{n} \sum_{k=1}^{n} A_{kj} - A_j\| \to 0$$

and

$$\left(\tfrac{1}{n} \sum_{k=1}^{n} \|A_{nj}\| \right) \text{ converges}, \ \sup_z \|D\psi_j(z)\| < \infty \ (j = 1, \dots, M).$$

As to the case $M = 1$ with more general weights in the recursion and a slightly weakened assumption on $(\|A_{n1}\|)$, see §2.

PROOF OF THEOREM 1.14: The argument of the proof of a result of Eckhaus (1985) and Sanchez-Palencia (1986) on averaged equations given in Sanders and Verhulst (1985), is modified in order to obtain convergence instead of approximation orders and to replace boundedness assumptions by ergodicity conditions and a local stability condition by a global one. At first one notices that $n_1 \in \mathbb{N}$, $E \in \{2, 3, \dots\}$, $k^* < 1$ exist such that for all sequences (y_n), (y_n^*) with

$$y_{n+1}^{(*)} = y_n^{(*)} - \tfrac{1}{n} h(y_n^{(*)}), \ n \in \mathbb{N},$$

the relations

$$\|y_n - y_n^*\| \le k^* \|y_m - y_m^*\|$$

for all $m, n \in \mathbb{N}$ with $m \ge n_1$, $n/m \ge E$ hold; further all these recursively defined sequences converge to ϑ. These results follow from the stability theory in Daleckii and Krein (1970/1974, section I.4), especially by use of the auxiliary norm $\| \ \|_{A,r}$ there, with large r, where $\|1 - \tfrac{A}{n}\|_{A,r} \le 1 - \tfrac{c}{n}$ for suitable $c > 0$ and sufficiently large n (compare also Walk and Zsidó 1989). Now one chooses $n^* \in \mathbb{N}$ so large that for arbitrary $c \in \mathbb{B}$ the equation $c = x - \tfrac{1}{n^*} h(x)$ has exactly one solution, and then $N' \in \{2, 3, \dots\}$ such that $E^{N'-1} \ge n^*$. Then for $N = N', N'+1, \dots$ the sequence $z^N := (z_n^N)_{n = E^{N-1}, E^{N-1}+1, \dots}$ is defined by

$$z_n^N = x_{E^N} \quad \text{for } n = E^N,$$
$$z_{n+1}^N = z_n^N - \tfrac{1}{n} h(z_n^N) \quad \text{for } n = E^{N-1}, \, E^{N-1} + 1, \ldots.$$

Further the sequence $z := (z_n)_{n \in \mathbb{N}}$ is defined by

$$z_1 = X_1,$$
$$z_{n+1} = z_n - \tfrac{1}{n} h(z_n).$$

For N sufficiently large, with $I_N := \{E^N, \ldots, E^{N+1}\}$ one obtains

$$\max_{k \in I_N} \| z_k - z_k^N \| =: \| z - z^N \|_{I_N} \leq k^* \| z - z^N \|_{I_{N-1}}$$

and, because h satisfies a Lipschitz condition,

$$\| z^N - z^{N-1} \|_{I_{N-1}} \leq k^{**} \| x_{E^N} - z_{E^N}^{N-1} \|$$

for suitable $k^{**} < \infty$, thus, with $x := (X_n)$ and obvious notation,

$$\| x - z \|_{I_N} \qquad \leq \| x - z^N \|_{I_N} + \| z - z^N \|_{I_N},$$
$$\| z - z^N \|_{I_N} \qquad \leq k^*(\| x - z \|_{I_{N-1}} + \| x - z^N \|_{I_{N-1}}),$$
$$\| x - z^N \|_{I_{N-1}} \leq \| x - z^{N-1} \|_{I_{N-1}} + k^{**} \| x - z^{N-1} \|_{I_{N-1}}.$$

Similarly to the proof of Theorem 3.3.3 in Sanders and Verhulst (1985), but using several times partial summation in the context of the ergodicity assumptions, one obtains existence of a monotone zero sequence (ε_N) in $\mathbb{R}_+$ with

$$\| x - z^N \|_{I_N} \leq \varepsilon_N + \varepsilon_N \| x \|_{I_N}, \quad N \in \mathbb{N}.$$

In view of this auxiliary result one argues in the following way, where $c' \in \mathbb{R}_+$ and the zero sequence (ε_N') in $\mathbb{R}_+$ are suitably chosen. In the first step for $N = 1, 2, \ldots$ and $2 \leq T = T(N) \leq E^N$ let

$$f_T(n, y) \quad := \quad \tfrac{1}{T} \sum_{i=1}^{T} \tfrac{1}{n+i} f(n+i, y), \quad y \in \mathbb{B},$$
$$y_{n+1}^N \quad := \quad y_n^N - f_T(n, y_n^N), \quad n \in I_N,$$

with

$$y_{E^N}^N := x_{E^N};$$

further let

$$\Phi(n, N) := \sum_{j=E^N}^{n} \tfrac{1}{j} f(j, \tilde{x}_j^N), \quad n \geq E^N,$$

with

$$\tilde{x}_j^N := x_j, \, j \in I_N, \, \tilde{x}_j^N := x_{E^{N+1}}, \, j > E^{N+1},$$

and

$$\Phi_T(n, N) := \tfrac{1}{T} \sum_{k=1}^{T} \Phi(n+k, N), \quad n \in I_N.$$

Thus

$$\| x_{n+1} - y_{n+1}^N \|$$
$$= \| \Phi(n, N) - \sum_{k=E^N}^{n} f_T(k, y_k^N) \|$$
$$\leq \| \Phi(n, N) - \Phi_T(n, N) \|$$
$$\quad + \| \Phi_T(n, N) - \sum_{k=E^N}^{n} f_T(k, x_k) \|$$

$$+ \sum_{k=E^N}^{n} \|f_T(k, x_k) - f_T(k, y_k^N)\|, \quad n \in I_N.$$

One obtains

$$\|\Phi(n, N) - \Phi_T(n, N)\|$$

$$= \|\frac{1}{T} \sum_{k=1}^{T} \sum_{j=n+1}^{n+k} \frac{1}{j} f(j, \tilde{x}_j^N)\|$$

$$\leq \|\frac{1}{T} \sum_{k=1}^{T} \sum_{j=n+1}^{n+k} \frac{1}{j} f(j, x_1)\| + \frac{1}{T} \sum_{k=1}^{T} \sum_{j=n+1}^{n+k} \frac{1}{j} g_j \|\tilde{x}_j^N - x_1\|$$

$$\leq (1 + \|x\|_{I_N})(\varepsilon_N' + c' \frac{T}{E^N}), \quad n \in I_N.$$

Further

$$\|\Phi_T(n, N) - \sum_{k=E^N}^{n} f_T(k, x_k)\|$$

$$\leq (1 + \|x\|_{I_N})(\varepsilon_N' \frac{E^N}{T} + c' \frac{T}{E^N}), \quad n \in I_N.$$

For the left side is majorized by

$$\|\frac{1}{T} \sum_{k=1}^{T} \sum_{j=E^N}^{E^N+k-1} \frac{1}{j} f(j, x_j)\|$$

$$+\|\frac{1}{T} \sum_{j=E^N}^{n} \sum_{k=1}^{T} \frac{1}{j+k} [f(j+k, \tilde{x}_{j+k}^N) - f(j+k, x_j)]\|;$$

here the first summand is treated as before; the second summand has

$$\frac{1}{T} \sum_{j=E^N}^{E^{N+1}} \sum_{k=1}^{T} \frac{1}{j+k} g_{j+k} \cdot \max_{j \in I_N, k \leq T} \|\tilde{x}_{j+k}^N - x_j\|$$

as an upper bound, where the first factor and

$$\|\tilde{x}_{j+k}^N - x_j\| = \| \sum_{l=j}^{\min\{j+k, E^{N+1}\}-1} \frac{1}{l} f(l, x_l)\|$$

can also be treated as before. Moreover

$$\sum_{k=E^N}^{n} \|f_T(k, x_k) - f_T(k, y_k^N)\|$$

$$\leq \frac{1}{T} \sum_{k=E^N}^{n} (\sum_{j=k+1}^{k+T} \frac{1}{j} g_j) \|x_k - y_k^N\|$$

$$\leq (\frac{c'}{E^N} + \frac{\varepsilon_N'}{T}) \sum_{k=E^N}^{n} \|x_k - y_k^N\|, \quad n \in I_N.$$

In the next step set

$$z_{n+1}^N := z_n^N - \frac{1}{n} h(z_n^N), \quad n \in I_N,$$

with

$$z_{E^N}^N := x_{E^N}.$$

Then, with Lipschitz continuity of h and $y_k^N = -(x_k - y_k^N) + x_k$, one obtains

$$\|y_{n+1}^N - z_{n+1}^N\|$$

$$= \| - \sum_{k=E^N}^{n} f_T(k, y_k^N) + \sum_{k=E^N}^{n} \tfrac{1}{k} h(z_k^N) \|$$

$$\leq \| \sum_{k=E^N}^{n} \tfrac{1}{k}[h(y_k^N) - h(z_k^N)] \|$$

$$+ T \sum_{k=E^N}^{n} \tfrac{1}{k^2} \| h(y_k^N) \| + \sum_{k=E^N}^{n} \tfrac{1}{T} \| \sum_{i=k+1}^{k+T} \tfrac{1}{i} [f(i, y_k^N) - h(y_k^N)] \|$$

$$\leq (\tfrac{c'}{E^N} + \tfrac{\varepsilon_N'}{T}) \sum_{k=E^N}^{n} (\| y_k^N - z_k^N \| + \| x_k - y_k^N \|)$$

$$+ (c' \tfrac{T}{E^N} + \tfrac{\varepsilon_N' E^N}{T}) \cdot (1 + \|x\|_{I_N}), \quad n \in I_N.$$

Now, by induction in view of the second inequality,

$$\| x_n - z_n^N \|$$
$$\leq \| x_n - y_n^N \| + \| y_n^N - z_n^N \|$$
$$\leq (1 + \|x\|_{I_N})(c \tfrac{T}{E^N} + \tfrac{\varepsilon_N E^N}{T}) e^{(c/E^N + \varepsilon_N/T)(n - E^N)}$$
$$\leq (1 + \|x\|_{I_N})(c \tfrac{T}{E^N} + \tfrac{\varepsilon_N E^N}{T}) e^{(c + \varepsilon_N E^N/T)(E-1)}, \quad n \in I_N,$$

with $c = 3c'$, $\varepsilon_N = 3\varepsilon_N'$, and a suitable choice of $T = T(N)$ yields the auxiliary result.

Now, with $\varepsilon_N^* := (1 + k^* + k^* k^{**})\varepsilon_N$ and N sufficiently large,

$$\| x - z \|_{I_{N+1}} \leq \varepsilon_N^* (1 + \max\{\|x\|_{I_N}, \|x\|_{I_{N+1}}\}) + k^* \| x - z \|_{I_N}$$
$$\leq \varepsilon_N^* (1 + \max\{\|x - z\|_{I_1}, \ldots, \|x - z\|_{I_{N+1}}\} + \sup_n \|z_n\|)$$
$$+ k^* \| x - z \|_{I_N}.$$

This yields

$$\| x - z \|_{I_N} \to 0,$$

from which $x_n \to \vartheta$ follows. $\qquad\qquad\qquad\qquad\qquad\qquad\qquad\square$

§2   Recursive methods for linear problems

In linear filtering and regression theory the following problem appears (see the cited monographs and Györfi 1984). Let Y be a real random variable, Z a k-dimensional random vector with $EY^2 < \infty$, $E\|Z\|^2 < \infty$, where realizations of Z can be observed. One seeks for an $x \in \mathbb{R}^k$ which for each realization (y, z) of (Y, Z) with observed z and unobserved y a linear estimate $< z, x >$ (scalar product) of y yields such that $E|Y - < Z, x >|^2$ is minimized. Such an x solves the equation

(1) $\qquad Ax - b = 0$

with $k \times k$-matrix $A = EZZ'$ and $b = EYZ \in \mathbb{R}^k$. Let A be positive definite. Then (1) has a unique solution which shall be denoted by ϑ. It is assumed that a training sequence of pairs (A_n, b_n) with $k \times k$-matrix valued random variables A_n and k-vector valued random variables b_n satisfying

(2) $\qquad \| \tfrac{1}{n}(A_1 + \ldots + A_n) - A \| \longrightarrow 0 \ (n \to \infty),$

(3) $\qquad \| \tfrac{1}{n}(b_1 + \ldots + b_n) - b \| \longrightarrow 0 \ (n \to \infty)$

a.s. is observable, e.g. via a training sequence of pairs (Y_n, Z_n) of the same type as (Y, Z) above with the corresponding ergodicity property. On the basis of successively observed (A_n, b_n) a recursive estimation of ϑ shall be given.

Analogous problems appear in the context of stochastic processes and lead to the case of $\mathbb{B}$-valued random variables X_n, b_n and $L(\mathbb{B})$-valued random variables A_n and elements $b \in \mathbb{B}$, $A \in L(\mathbb{B})$, where $\mathbb{B}$ is a real separable Banach space. As to (1) in $\mathbb{B} = C([0, 1]^2)$ provided with max-norm, see Arnold (1973, p. 219). The assumption

(4) $\qquad spec\,(A) \subset \{\lambda \in \mathbb{C};\ re\,\lambda > 0\}$

guarantees that (1) possesses a unique solution $\vartheta = A^{-1}b$. This situation will be considered in the following. In deterministic considerations the assumption of separability of $\mathbb{B}$ can be dropped.

Let a sequence (X_n) of $\mathbb{B}$-valued random variables be defined by a Widrow algorithm

(5) $\qquad X_{n+1} = X_n - \frac{1}{n}(A_n X_n - b_n)$

(Widrow and Hoff, Jr., 1960) under the assumptions (2), (3) and

(6) $\qquad \frac{1}{n}(\|A_1\| + \ldots + \|A_n\|) = O(1)$

a.s. together with spectral condition (4) on A. In contrast to the classical Robbins-Monro situation the error terms do not appear only additively. The following theorem (Walk and Zsidó 1989, with further references to literature, e.g. Györfi 1980) which concerns more general weights and has a purely deterministic formulation, yields a.s. convergence $X_n \to \vartheta$ $(n \to \infty)$.

2.1. Theorem. *Let $b \in \mathbb{B}$, $A \in L(\mathbb{B})$ with (4), $a_n \in [0, 1)$ $(n \in \mathbb{N})$ with $a_n \to 0$ $(n \to \infty)$, $\sum a_n = \infty$, and β_n, γ_n according to Lemma 1.1. For $A_n \in L(\mathbb{B})$, $b_n \in \mathbb{B}$, $X_n \in \mathbb{B}$ $(n \in \mathbb{N})$ assume*

(2′) $\qquad \left\| \beta_n^{-1} \sum\limits_{k=1}^{n} \gamma_k A_k - A \right\| \to 0,$

(6′) $\qquad \beta_n^{-1} \sum\limits_{k=1}^{n} \gamma_k \|A_k\| = O(1),$

(3′) $\qquad \left\| \beta_n^{-1} \sum\limits_{k=1}^{n} \gamma_k b_k - b \right\| \to 0,$

(5′) $\qquad X_{n+1} = X_n - a_n(A_n X_n - b_n).$

Then $\qquad\qquad X_n \to \vartheta\ (= A^{-1}b).$

2.2. Remark on Theorem 2.1. a) In the case of a Euclidean space with symmetric positive semidefinite A_n, condition (6′) is implied by the other conditions.

b) In the case $a_n = \rho_n/n$ with

$$0 < \underline{\lim}\,\rho_n \le \overline{\lim}\,\rho_n < \infty,\ \rho_n - \rho_{n-1} = O(\tfrac{1}{n})$$

or with

$$\rho_n \to \rho > 0,\ \sum |\rho_n - \rho_{n-1}| < \infty$$

the equivalence relations

$$(2) \Longleftrightarrow (2'), \quad (3) \Longleftrightarrow (3'), \quad (6) \Longleftrightarrow (6')$$

hold.

In econometrics linear models with forecast feedback were treated in the last years. An example may be qualitatively described in the following way. Customers estimate the present development of prices in a more or less rational way by the development of prices in the past. But by purchase decisions these estimates influence the development of prices and thus also of estimates in the future. In other words: the present estimate is influenced also by estimates in the past. One speaks here of learning, especially in linear models, with forecast feedback. Kottmann 1990 and Mohr 1990, who give further references, investigate the behaviour of the estimations by use of stochastic approximation. There appears a linear recursion of the following form

$$X_{n+1} = \tfrac{1}{n} \sum_{j=1}^{n} B_j X_j + \tfrac{1}{n} \sum_{j=1}^{n} b_j,$$

where X_j, b_j are k-dimensional random vectors or more generally $\mathbb{B}$-valued random variables, B_j are random $k \times k$-matrices or more generally $L(\mathbb{B})$-valued random variables. With $A_n = 1 - B_n$ one obtains the above recursion with weights $a_n = \tfrac{1}{n}$, i.e.

$$X_{n+1} = X_n - \tfrac{1}{n}(A_n X_n - b_n),$$

where $X_1 = 0$. It should be noted that, in the k-dimensional case, the matrices are generally not symmetric, differently from the example at the beginning of this section.

At learning in linear models, also the case appears that an initial part of the past is forgotten, so-called learning with forgetting. Then the more general recursion

$$(7) \quad X_{n+1} = \tfrac{1}{n} \sum_{j=j_n}^{n} B_j X_j + d_n$$

appears, where

$$d_n = \tfrac{1}{n} \sum_{j=j_n}^{n} b_j.$$

As to its a.s. behaviour, in a pathwise formulation, the following generalization of Theorem 2.1 for $a_n = 1/n$ (and $X_1 = 0$, $b = 0$ without loss of generality, there $j_n = 1$, thus $\alpha = 0$) holds. It answers a question put by Kottmann.

2.3. Theorem. *Let $B \in L(\mathbb{B})$, $\alpha \in [0, 1)$ with*

$$(8) \quad spec(B) \subset \{z(1 - \alpha^z)^{-1}; \ z \in \mathbb{C}, \ re\, z \geq 1\}^c.$$

Assume $B_n \in L(\mathbb{B})$, $d_n \in \mathbb{B}$, $X_n \in \mathbb{B}$ with (4),

$$(9) \quad \|\tfrac{1}{n}(B_1 + \ldots + B_n) - B\| \to 0,$$

$$(10) \quad \tfrac{1}{n}(\|B_1\| + \ldots + \|B_n\|) = O(1),$$

$$(11) \quad d_n \to 0,$$

where $j_n \in \{1, \ldots, n\}$, $j_n/n \to \alpha$. Then

$X_n \to 0$.

2.4. Remark. a) If in Theorem 2.3 the limit in (11) is $d \in \mathbb{B}$ instead of 0, then $X_n \to \vartheta$ with $\vartheta = (1 - (1 - \alpha)B)^{-1} d$.

b) Sufficient conditions for the spectral condition (8) are

$$\text{spec}(B) \subset \{w \in \mathbb{C}; \ |w| < \tfrac{1}{1-\alpha}\}$$

and

$$\text{spec}(B) \subset \{w \in \mathbb{C}; \ re\, w < 1 - |w|\alpha\}.$$

c) (8) can be considered as a stability condition. In the simplified recursion

$$x_{n+1} = \tfrac{b}{n} \sum_{j=[\alpha n]}^{n} x_k \ \text{ in } \mathbb{C}$$

and in the linear functional differential equation

$$\dot{x}(t) = -\tfrac{1}{t}x(t) + \tfrac{b}{t}x(t) - \tfrac{b}{t}x(\alpha t)\alpha \ \text{ in } \mathbb{C}$$

or in the corresponding retarded differential equation obtained by the transformation $u = \ln t$ (see Hale 1977, ch. 1) it appears in the form

$$b \in \{z(1 - \alpha^z)^{-1}; \ z \in \mathbb{C}, \ re\, z \geq 1\}^c$$

and yields $x_n \to 0 \ (n \to \infty)$ and $x(t) \to 0 \ (t \to \infty)$, resp. Both of the sufficient conditions in b) can be obtained via the above equations by a contraction principle or by a general result of Becker and Greiner (1986) on functional differential equations. It should be mentioned that the stability condition also appears in summability theory, namely in Mercer theorems for a modified Cesàro method where the first $[\alpha n]$ terms in the arithmetic means are cancelled.

SKETCH OF THE PROOF OF THEOREM 2.3: According to the averaging principle one writes recursion (7) in the form

$$X_{n+1} = \tfrac{B}{n} \sum_{j=j_n}^{n} X_j + \tfrac{1}{n} \sum_{j=j_n}^{n} (B_j - B)X_j + d_n.$$

Now the last sum is treated by partial summation, by use of (7) for $X_j - X_{j-1}$, and once more by partial summation. Then one obtains the representation

$$X_{n+1} = \tfrac{B}{n} \sum_{j=j_n}^{n} X_j + h_{n+1},$$

where $\|h_{n+1}\| \leq \varepsilon_n + \varepsilon_n \max\{\|X_1\|, \ldots, \|X_n\|\}$ with $\varepsilon_n \to 0$.

Imbedding into an integral equation

$$X(t) = \tfrac{B}{t} \int_{\alpha t}^{t} X(s)ds + g(t),$$

where $\|g(t)\| \leq \varepsilon(t) + \varepsilon(t) \sup_{2 \leq s \leq t} \|X(s)\|$ with $\varepsilon(t) \to 0 \ (t \to \infty)$,

and setting

$$Y(t) := \int_{1}^{t} X(s)ds$$

yield

$Y(t) = X(t) = \frac{B}{t}[Y(t) - Y(\alpha t)] + g(t),\ t \geq 2,$

where $Y(\alpha t) \equiv 0$ for $\alpha = 0$. The substitution $u = \ln t$, $Z(u) = Y(t)$ leads to a linear functional differential equation which can be treated according to Hale (1977, ch. 1). One obtains

$Z(u)/e^u \to 0,\ Y(t)/t \to 0,\ X(t) \to 0,\ X_n \to 0.$ □

As is well known, for an infinite-dimensional Hilbert space $\mathbb{H}$ a compact symmetric positive-semidefinite operator $A \in L(\mathbb{H})$ has a set of eigenvalues with 0 as an accumulation point. For this reason results in the case that the spectral condition (4) is not fulfilled, are of interest. The following theorem somewhat generalizes a result of Shwartz and Berman (1989) concerning weak convergence of a deterministic, i.e. pathwise considered, sequence of estimates in $\mathbb{H}$.

2.5. Theorem. *Assume $b \in \mathbb{H}$, further $A \in L(\mathbb{H})$ symmetric positive-semidefinite such that $Ax - b = 0$ has exactly one solution $\vartheta \in \mathbb{H}$. Let $a_n \in [0,1)$ with $a_n \to 0$, $\sum a_n = \infty$ and choose β_n, γ_n ($n \in \mathbb{N}$) according to Lemma 1.1. Assume $A_n \in L(\mathbb{H})$ ($n \in \mathbb{N}$) symmetric positive-semidefinite with $(2')$, $(6')$ and $b_n \in \mathbb{H}$ ($n \in \mathbb{N}$) with $(3')$. Further assume*

$$(12) \qquad \sum \alpha_{n+1} \left\| (1 - A_{n+1}) \frac{1}{\beta_n} \sum_{k=1}^{n} \gamma_k (A_k \vartheta - b_k) \right\| < \infty.$$

For (X_n) in $\mathbb{H}$ with $(5')$, it holds

$(13) \qquad X_n \to \vartheta$ *weakly,*

$(14) \qquad (\|X_n\|)$ *convergent,* $(\|X_n - \vartheta\|)$ *convergent,*

further, with $\|x\|_A := (x, Ax)^{1/2}$, $x \in \mathbb{H}$,

$(15) \qquad \|X_n - \vartheta\|_A \to 0,\ \|A(X_n - \vartheta)\| \to 0.$

PROOF: In view of (13), (15) and the second part of (14), without loss of generality, $b = 0$ and thus $\vartheta = 0$ can be assumed. Setting

$$e_n := \frac{1}{\beta_n} \sum_{k=1}^{n} \gamma_k b_k$$

one obtains (compare Walk and Zsidó 1989)

$$(16) \qquad X_{n+1} = e_n + \sum_{k=1}^{n-1} a_{k+1}(1 - a_n A_n) \dots (1 - a_{k+2} A_{k+2})(1 - A_{k+1}) e_k$$
$$+ (1 - a_n A_n) \dots (1 - a_1 A_1) x_1.$$

$(2')$ yields $a_n \|A_n\| \to 0$ and, because A_n is positive-semidefinite,

$\|1 - a_n A_n\| \leq 1$ for sufficiently large n.

Then together with (12) one obtains boundedness of $(\|X_n\|)$. According to the averaging principle one writes

$X_{n+1} = X_n - a_n A X_n + a_n r_n$

with $r_n = -(A_n - A)x_n + b_n$ and notices

$$\|\tfrac{1}{\beta_n} \sum_{k=1}^{n} \gamma_k r_k\| \to 0.$$

The latter follows by partial summation, (5'), once more partial summation and use of boundedness of $(\|X_n\|)$, (2'), (6'), (3'). Now with

$$F(x) := \frac{1}{2}(x,\, Ax) \geq 0,\ x \in \mathbb{H},$$

Theorem 1.2b together with Remark 1.3c can be applied and yields

$$\|DF(X_n)\| = \|AX_n\| \to 0$$

and, by boundedness of (X_n),

$$(X_n,\, AX_n) \to 0.$$

From each of these results together with boundedness of (X_n), by spectral theory (Yosida 1968, XI, 5),

$$X_n \to 0 \text{ weakly}$$

follows. Now in (16) for sufficiently large $\mathbb{N} < n$ one separates

$$\sum_{k=1}^{n} = \sum_{k=N-1}^{n-1} + \sum_{k=1}^{N-2},$$

uses (3'), (12) and the fact that

$$(\|(1 - a_n A_n)\ldots(1 - a_1 A_1)y\|)$$

converges for all $y \in \mathbb{H}$ (proved by taking squares and noticing $a_n\|A_n\| \to 0$), and thus obtains convergence of $(\|X_n\|)$. Finally one considers the case of a general ϑ. It remains to show the first part of (14). But this follows from

$$\|X_n\|^2 = \|X_n - \vartheta\|^2 + 2(X_n - \vartheta,\, \vartheta) + \|\vartheta\|^2$$

together with (13) and the second part of (14).   □

2.6. Remark. If in Theorem 2.5 $\|a_i A_i\| < 1$ for all i, then

$$\inf_{X_1}\ \lim_{n}\ \|X_n - \vartheta\| = 0,$$

$$\inf_{X_1}\ \lim_{n}\ \|X_n\| = \|\vartheta\|,$$

where X_n of course depends on the starting point X_1. This follows by the argument used in the last two steps of the foregoing proof.

§ 3   Stochastic optimization under stochastic constraints

This section concerns recursive estimation of a minimal point of a real-valued function of k real variables under constraints, where the values of the objective function as well as the values of the functions describing the constraints are contaminated by random noise.

At first for the deterministic optimization problem where the function values are not contaminated, an approach due to Rockafellar (1973) is described. Let continuous functions $f_0, f_1, \ldots, f_m : \mathbb{R}^k \to \mathbb{R}$ be given. The primal problem $(\mathcal{P})$ is to minimize $\{f_0(x);\ x \in E\}$, where $E := \{x \in \mathbb{R}^k;\ f_i(x) \leq 0\ (i =$

$1, \ldots, m)\}$. For its treatment so-called penalty Lagrangians $L_0 : \mathbb{R}^k \times \mathbb{R}^m \to \overline{\mathbb{R}}$, $L_r : \mathbb{R}^k \times \mathbb{R}^m \to \overline{\mathbb{R}}$, $r > 0$, with $\overline{\mathbb{R}} = \mathbb{R} \cup \{+\infty, -\infty\}$ are introduced by

$$L_0(x, y) := \begin{cases} f_0(x) + \sum\limits_{j=1}^{m} y_j f_j(x) & , \text{ if } \underset{i \in \{1, \ldots, m\}}{\forall} \quad y_i \geq 0 \\ -\infty & , \text{ if } \underset{i \in \{1, \ldots, m\}}{\exists} \quad y_i < 0, \end{cases}$$

$$L_r(x, y) := f_0(x) + \tfrac{1}{4r} \sum\limits_{j=1}^{m} ([(y_j + 2r f_j(x))^+]^2 - y_j^2),$$

the latter in view of differentiability properties. One has

$$L_r(x, y) \to L_0(x, y) \quad (r \downarrow 0).$$

The dual problem $(\mathcal{D}_r)$ concerns

$$\sup_{y \in \mathbb{R}^m} \inf_{x \in \mathbb{R}^k} L_r(x, y),$$

where y is the so-called dual variable.

From now on, in the deterministic as well as in the stochastic case, the following assumptions shall be satisfied:

(1) $f_0, f_1, \ldots, f_m : \mathbb{R}^k \to \mathbb{R}$ convex;

(2) f_0 inf-compact, i.e. $\{x \in \mathbb{R}^k;\ f_0(x) \leq \lambda\}$ compact for each $\lambda \in \mathbb{R}$;

(3) $\underset{x_0 \in \mathbb{R}^k}{\exists} \ \underset{j \in \{1, \ldots, m\}}{\forall} \quad f_j(x_0) < 0$ (Slater condition).

Then the following deterministic results hold. The set $S (\subset \mathbb{R}^k)$ of optimal solutions of $\mathcal{P}$ is nonvoid convex and compact.

$Y := \{\overline{y} \in \mathbb{R}^m;\ -\infty < \min\limits_{x \in \mathbb{R}^k} L_r(x, \overline{y}) = (\min \text{ in } \mathcal{P})\}$, the set of Kuhn-Tucker vectors for L_r, is independent of $r \geq 0$ and is a nonvoid convex and compact subset of $\mathbb{R}^m_+$.

$Y = \{\overline{y} \in \mathbb{R}^m;\ \overline{y}$ optimal solution of $(\mathcal{D}_r)\}$.

The optimal value of $\mathcal{D}_r$ is independent of r.

L_r $(r > 0)$ is convex with respect to its first argument and concave with respect to its second argument. It holds for $r \geq 0$:

$(\overline{x}, \overline{y})$ is saddle-point of L_r

i.e. $\underset{x \in \mathbb{R}^k}{\forall} \ \underset{y \in \mathbb{R}^m}{\forall} \quad L_r(\overline{x}, y) \leq L_r(\overline{x}, \overline{y}) \leq L_r(x, \overline{y})$

$\Longleftrightarrow$ Kuhn-Tucker condition is fulfilled,

i.e. $\begin{cases} \underset{i \in \{1, \ldots, m\}}{\forall} \quad \overline{y}_i \geq 0,\ f_i(\overline{x}) \leq 0,\ \overline{y}_i f_i(\overline{x}) = 0 \\ \overline{x} \text{ minimizes } L_0(\cdot, \overline{y}) = f_0 + \sum\limits_{j=1}^{m} \overline{y}_j f_j \end{cases}$

$\Longleftrightarrow (\overline{x} \in S) \wedge (\overline{y} \in Y).$

$\underset{\overline{y} \in Y}{\forall} \ \underset{r > 0}{\forall} \quad [\overline{x} \in S \Longleftrightarrow \overline{x} \text{ minimizes } L_r(\cdot, \overline{y})].$

$$\underset{\overline{y}\in Y}{\forall} \qquad [\overline{x}\in S \Longleftarrow \overline{x} \text{ minimizes } L_0(\cdot,\overline{y})].$$

$$\underset{r>0}{\forall}\ \underset{i\in\{1,\dots,m\}}{\forall}\quad (D_y L_r(x,y))_i = \max\{f_i(x), -\tfrac{y_i}{2r}\},$$

$$\underset{r>0}{\forall}\quad D_x L_r(x,y) = Df_0(x) + \sum_{j=1}^{m}[y_j + 2rf_j(x)]^+ Df_j(x),$$

if $f_0, f_1,\dots,f_m$ are totally differentiable.

The basic idea for the deterministic and stochastic case to find a maximal point $\overline{y}\in Y$ of $\underset{x\in\mathbb{R}^k}{\inf}\ L_r(x,y)$ and then to find a minimal point $\overline{x}\in\mathbb{R}^k$ of $L_r(\cdot,\overline{y})$ as an optimal solution of $(\mathcal{P})$ leads, under differentiability assumptions, to a stepwise procedure with the aim to approximate a solution of

$$D_y L_r(x,y) \equiv (\max\{f_i(x), -\tfrac{y_i}{2r}\})_{i=1,\dots,m} = 0,$$
$$D_x L_r(x,y) = 0.$$

This stepwise procedure will be achieved in the stochastic case by successive pairs consisting of a Robbins-Monro type step (with changed sign of the correction term because of sup instead of inf) and a Kiefer-Wolfowitz type step. It has been proposed and investigated by Kushner and Sanvicente (1975) and Hiriart-Urruty (1977), see also Kushner and Clark (1978). The advantage of this primal-dual method in the stochastic case is that the subproblems are optimization problems without constraints. Thus one avoids the decision whether an $x\in\mathbb{R}^k$ belongs to E or E^c, which is important in various other optimization procedures, but not possible in the stochastic case because of the observation errors.

The following regularity assumptions shall be fulfilled:

(4) $f_0,\dots,f_m$ are continuously differentiable, where Df_i is bounded
       $(i=1,\dots,m)$ and Df_i satisfies a Lipschitz condition $(i=0,\dots,m)$;

(5) $L_0(\overline{x},\overline{y}) < L_0(x,\overline{y})$ for all $x\in\mathbb{R}^k$ with projection (on S) $\overline{x}\neq x$ and all
       Kuhn-Tucker vectors $\overline{y}$;

(6) $\underset{x\in\mathbb{R}^k}{\inf}\ f_i(x) > -\infty\ (i=1,\dots,m)$.

In examples the values $f_i(x)$ $(i=1,\dots,m)$ in the constraints are often probability values. If in the algorithm below $\rho_n = 0$ is chosen (as Kushner and Sanvicente did), (6) can be cancelled.

The recursion sequence consists of pairs (X_n, Y_n) where X_n are k-dimensional vectors, $Y_n = (Y_{ni})_{i=1,\dots,m}$ are m-dimensional random vectors with $Y_{ni}\geq 0$ $(n\in\mathbb{N})$. Let a_n, c_n be positive real numbers, ρ_n be nonnegative real numbers $(n\in\mathbb{N})$ with

$$c_n \to 0,\quad \sup_n \rho_n < \infty,$$

$$\sum a_n^2 c_n^{-2} < \infty,\quad \sum a_n = \infty,\quad \sum a_n c_n < \infty,\quad \sum a_n \rho_n c_n^{-1} < \infty.$$

With square integrable random variables V_n^i $(i=1,\dots,m;\ n\in\mathbb{N})$ and V_{nl}^i $(i=0,\dots,m;\ l=1,\dots,k;\ n\in\mathbb{N})$ which describe the contamination of function values, the recursion for (X_n, Y_n) is given by

(7') $X_{n+1} = X_n - a_n \tilde{D}_x L_{\rho_n}(X_n, Y_n)$,

(7'') $Y_{n+1} = Y_n + a_n \tilde{D}_y L_{a_n/2}(X_n, Y_n)$.

Here

$$(\tilde{D}_x L_{\rho_n}(X_n, Y_n))_l$$
$$= (2c_n)^{-1}[f_0(X_n + c_n e_l) - f_0(X_n - c_n e_l) - V_{nl}^0]$$
$$+ \sum_{j=1}^{m}[Y_{nj} + 2\rho_n(f_j(X_n) - V_n^j)]^+ (2c_n)^{-1}[f_j(X_n + c_n e_l) - f_j(X_n - c_n e_l) - V_{nl}^j]$$

(e_l k-dimensional unit vector with 1 as l-th coordinate; $l = 1, \ldots, k$),

$$(\tilde{D}_y L_{a_n/2}(X_n, Y_n))_i$$
$$= \max\{f_i(X_n) - V_n^i, -a_n^{-1} Y_{ni}\} \ (i = 1, \ldots, m).$$

With σ-algebras $\mathcal{A}_n$ generated by $X_1, Y_1, V_1^i, \ldots, V_{n-1}^i$ ($i = 1, \ldots, m$), $V_{1l}^i, \ldots, V_{n-1,l}^i$ ($i = 0, \ldots, m$; $l = 1, \ldots, k$), it is assumed

(8) $\forall_{n,l,i} \quad E(V_{nl}^i | \mathcal{A}_n) = 0, \ E(V_n^i | \mathcal{A}_n) = 0,$

(9) $\forall_{n,l} \quad E((V_{nl}^0)^2 | \mathcal{A}_n) \leq$ const ,

(10) $\forall_{n,l} \quad \forall_{i \in \{1, \ldots, m\}} \quad E((V_n^i)^4 | \mathcal{A}_n) + E((V_{nl}^i)^4 | \mathcal{A}_n) \leq$ const

for some const $< \infty$.

3.1. Theorem. *Let the sequence of random elements (X_n, Y_n) in $\mathbb{R}^k \times \mathbb{R}_+^m$ be recursively defined by (7'), (7'') and assume (1) – (6), (8) – (10). Then*

$$\text{dist}\,(X_n, S) \to 0 \text{ a.s.}$$

also, if X_1, Y_1 are square integrable, in first mean.

3.2. Remark. a) If f_0 is strictly convex, then $L_0(\cdot, \bar{y})$ has exactly one minimal point ϑ, thus $S = \{\vartheta\}$ and (5) is fulfilled.
b) In the case that no constraints are given, Theorem 3.1 is a result on the Kiefer-Wolfowitz process.

SKETCH OF THE PROOF OF THEOREM 3.1: (Compare the proof of Theorem 1.9.) The set of saddle-points of L_0 or L_r is $S \times Y$; it is convex and compact. Let $(\overline{X}_n, \overline{Y}_n)$ be the – unique – projection of (X_n, Y_n) onto $S \times Y$. $\overline{X}_n, \overline{Y}_n$ are the unique projections of X_n onto S and of Y_n onto Y, resp. Without loss of generality, X_1 and Y_1 are assumed square integrable. In the first step one shows

$$E(\|Y_{n+1} - \overline{Y}_{n+1}\|^2 | \mathcal{A}_n)$$
$$\leq \|Y_n - \overline{Y}_n\|^2 + 2a_n(D_y L_0(X_n, Y_n), Y_n - \overline{Y}_n) + \text{const.} \ a_n^2 + \text{const.} \ a_n^2 \|X_n - \overline{X}_n\|^2.$$

In the second step one shows

$$E(\|X_{n+1} - \overline{X}_{n+1}\|^2 | \mathcal{A}_n)$$
$$\leq (1 + \alpha_n')\|X_n - \overline{X}_n\|^2 + \beta_n'\|Y_n - \overline{Y}_n\|^2 + \delta_n' - 2a_n(D_x L_0(X_n, Y_n), X_n - \overline{X}_n),$$

where α_n', β_n', δ_n' are positive real numbers with

$$\sum a'_n < \infty, \quad \sum \beta'_n < \infty, \quad \sum \delta'_n < \infty.$$

In the third step one considers the squared distance

$$A_n := \|X_n - \overline{X}_n\|^2 + \|Y_n - \overline{Y}_n\|^2,$$

further

$$B_n := (D_x L_0(X_n, Y_n), X_n - \overline{X}_n) - (D_y L_0(X_n, Y_n), Y_n - \overline{Y}_n).$$

Because L_0 is convex-concave on $\mathbb{R}^k \times \mathbb{R}^m_+$ and $(\overline{X}_n, \overline{Y}_n)$ is a (random) element in the set $S \times Y$ of saddle-points of L_0,

$$B_n \geq [\, L_0(X_n, Y_n) - L_0(\overline{X}_n, Y_n)\,] - [\, L_0(X_n, Y_n) - L_0(X_n, \overline{Y}_n)\,] \geq 0.$$

The inequalities of the first two steps yield

$$E(A_{n+1}|\mathcal{A}_n) \leq (1 + \gamma_n) A_n - 2 a_n B_n + \gamma_n$$

with positive real numbers γ_n satisfying $\sum \gamma_n < \infty$. Thus (A_n) is a nonnegative almost supermartingale. As in the proof of Theorem 1.9 it follows, by Lemma 1.10:

(A_n) converges a.s., (EA_n) converges,

further $\sum a_n EB_n < \infty$.

Now $\|X_n - \overline{X}_n\| \to 0$ a.s., i.e. dist $(X_n, S) \to 0$ a.s., is indirectly proved by a pathwise consideration neglecting a set of probability measure zero (compare the argument in the proof of Theorem 1.1a). At first it is noticed that for each $\varepsilon > 0$, by (5), a $\delta(\varepsilon)$ exists such that $B_j \geq \delta(\varepsilon)$ for all j with $\|X_j - \overline{X}_j\| \geq \varepsilon/2$. Assume now existence of an $\varepsilon > 0$ with $\|X_l - \overline{X}_l\| \geq \varepsilon$ for infinitely many l. The argument in the second step together with a martingale argument yields

$$X_{n+1} = X_n - a_n D_x L_0(X_n, Y_n) + D_n$$

with convergence of $\sum D_n$. Because of boundedness of $(\|X_n\|)$ and $(\|Y_n\|)$,

$$K := \sup_n \|D_x L_0(X_n, Y_n)\| < \infty.$$

Now an N is chosen such that $\|X_N - \overline{X}_N\| \geq \varepsilon$,

$$\sum_{j=N}^{\infty} a_j B_j \leq \tfrac{1}{K} \delta(\varepsilon) \tfrac{\varepsilon}{8}, \quad \forall_{n \geq N} \quad \|\sum_{j=n}^{\infty} D_j\| \leq \tfrac{\varepsilon}{8}.$$

From the induction assumption that $\|X_j - \overline{X}_j\| \geq \varepsilon/2$ for $j = N, \ldots, n$, it follows

$$\|X_{n+1} - X_N\| \leq \|\sum_{j=N}^{n} D_j\| + K \sum_{j=N}^{n} a_j B_j / \delta(\varepsilon) \leq \tfrac{\varepsilon}{4},$$

thus

$$\|\overline{X}_{n+1} - \overline{X}_N\| \leq \tfrac{\varepsilon}{4} \text{ and } \|X_{n+1} - \overline{X}_{n+1}\| \geq \tfrac{\varepsilon}{2}.$$

But $\|X_n - \overline{X}_n\| \geq \tfrac{\varepsilon}{2}$, $n \geq \mathbb{N}$, implies $B_n \geq \delta(\varepsilon)$, $n \geq N$, which is in contrast to $\sum a_n B_n < \infty$ and $\sum a_n = \infty$.

In the fourth step one notices $Ed(X_n, S)^2 \leq EA_n = O(1)$, by the second step, thus uniform integrability of $(d(X_n, S))$ and, because of the third step, $Ed(X_n, S) \to 0$. $\qquad \square$

3.3. Remark. If in Theorem 3.1 $\rho_n = 0$ $(n \in \mathbb{N})$ is chosen, conditions (9) and (10) can be weakened to

$$\underset{l,i}{\forall} \quad \sum a_n^2 c_n^{-2} E(V_{nl}^i)^2 < \infty,$$

$$\underset{i}{\forall} \quad \sum a_n^2 E(V_n^i)^2 < \infty.$$

3.4. Remark. The assertion of a.s. convergence in Theorem 3.1 and Remark 3.3 can be sharpened to
a.s. convergence of (X_n) to a random element in S,
a.s. $\|Y_n - \overline{Y}_n\| \to 0$
(Walk 1983-84). For the proof of the first part compare the proof of Theorem 1.11b and Pflug (1981). The proof of the second part uses summability theory.

§4   A learning model; recursive density estimation

In this section a special situation of §1 is treated. It concerns learning theory with an application to density estimation.

In the first part the limit of expectations of observable k-dimensional random vectors Z_n shall be recursively estimated by the learning rule

(1) $X_{n+1} = (1 - a_n)X_n + a_n Z_n, \ n \in \mathbb{N},$

with $a_n \in [0, 1)$ and k-dimensional random vectors X_n (see Pakes 1982 and the literature cited there). Analogously the case of a real separable Hilbert space instead of $\mathbb{R}^k$ can be treated.

4.1. Theorem. *Let X_n, Z_n $(n \in \mathbb{N})$ be k-dimensional random vectors with square integrability of Z_n, further $\vartheta \in \mathbb{R}^k$ and $a_n \in [0, 1)$ $(n \in \mathbb{N})$ with $a_n \to 0$, $\sum a_n = \infty$ such that (1),*
(2) $E(Z_n | Z_1, \dots, Z_{n-1}) = EZ_n \ (n = 2, 3, \dots), \ \sum a_n^2 E\|Z_n\|^2 < \infty,$
(3) $EZ_n \to \vartheta$
hold. Then
$$X_n \to \vartheta \ a.s.$$

PROOF: From (2), by martingale theory, Lemma 1.1 with its notations and the Kronecker lemma, it follows

(4) $\frac{1}{\beta_n} \sum\limits_{j=1}^{n} \gamma_j (Z_j - EZ_j) \to 0$ a.s. ,

which could also be assumed instead of (2). Lemma 1.1 yields that the representations (1) and

$$X_{n+1} = \frac{1}{\beta_n} \sum_{j=1}^{n} \gamma_j Z_j + \prod_{j=1}^{n} (1 - a_j) X_1, \ n \in \mathbb{N},$$

are equivalent, and further

$$\frac{1}{\beta_n} \sum_{j=1}^{n} \gamma_j EZ_j \to \vartheta$$

because of (3). From these statements the assertion follows. – Another proof notices that (1) is the recursion of Theorems 1.2 and 1.9 with $F(x) = \frac{1}{2}\|x - \vartheta\|^2$

and $f(x) = x - \vartheta$, resp., where $H_n = EZ_n - \vartheta \to 0$, $V_n = Z_n - EZ_n$ with (4). Theorem 1.2b then yields $DF(X_n) \to 0$ a.s., i.e. the assertion. □

The second part of this section concerns recursive density estimation. Assume that the independent identically distributed (i.i.d.) real random variables $Y_1, Y_2, \ldots$ with observable realizations $y_1, y_2, \ldots$ possess a bounded density $f : \mathbb{R} \to \mathbb{R}_+$. Let the so-called kernel function $K : \mathbb{R} \to \mathbb{R}_+$ be a fixed bounded density function, e.g. of uniform distribution on $[-1/2, 1/2]$ or standard normal distribution, and (h_n) be a sequence of positive real numbers with $h_n \to 0$ (so-called window widths). The Rosenblatt-Parzen estimate (1956/1962) of $f(y)$, $y \in \mathbb{R}$, on the basis of observed $y_1, \ldots, y_n$ is given by

$$f_n(y_1, \ldots, y_n, y) := \frac{1}{nh_n} \sum_{j=1}^{n} K\left(\frac{y-y_j}{h_n}\right), \quad n \in \mathbb{N}.$$

Such an estimate is motivated by the idea to smooth the unknown density f by $\tilde{f}_n$ where

$$\tilde{f}_n(y) = \frac{1}{h_n} \int_{\mathbb{R}} K\left(\frac{y-t}{h_n}\right) f(t)dt$$

and to notice that an unbiased estimate of this singular integral is given by the above $f_n(\cdot, \ldots, \cdot, y)$. Smoothing leads to a systematic error, unbiased estimation to a stochastic error.

The following lemma, here in a simple form, is well-known in approximation theory and shall later on be used in the treatment of errors.

4.2. Lemma. *Let $f : \mathbb{R} \to \mathbb{R}_+$ be a bounded density and $K^* : \mathbb{R} \to \mathbb{R}_+$ be integrable. The singular integrals*

$$g_v(y) := \frac{1}{v} \int_{\mathbb{R}} K^*\left(\frac{y-t}{v}\right) f(t)dt \quad (y \in \mathbb{R}, \ v > 0)$$

for f with kernel K^* have the property that for each continuity point $y \in \mathbb{R}$ of f the relation

$$g_v(y) \to f(y) \int_{\mathbb{R}} K^*(t)dt \quad (v \to 0)$$

holds.

PROOF: Either by a separate treatment of f on $[y-\delta, y+\delta]$ and on $[y-\delta, y+\delta]^c$ with small $\delta > 0$ or by a substitution $u = (y-t)/v$ together with the dominated convergence theorem. □

A recursive variant of the Rosenblatt-Parzen estimation sequence with K and (h_n) as before, but with more general weights, is defined on the basis of successively observed $y_1, y_2, \ldots$ by

$$\hat{f}_1(y_1, y) := \frac{1}{h_1} K\left(\frac{y-y_1}{h_1}\right),$$

$$\hat{f}_n(y_1, \ldots, y_n, y) := (1 - a_n)\hat{f}_{n-1}(y_1, \ldots, y_{n-1}, y) + a_n \frac{1}{h_n} K\left(\frac{y-y_n}{h_n}\right)$$

$(n = 2, 3, \ldots)$ with $a_n \in [0, 1)$, $a_n \to 0$, $\sum a_n = \infty$. A non-recursive representation is given by

$$\hat{f}_n(y_1, \ldots, y_n, y) = \frac{1}{\hat{\beta}_n} \sum_{j=1}^n \hat{\gamma}_j \frac{1}{h_j} K\left(\frac{y - y_j}{h_j}\right),$$

where

$$\hat{\gamma}_1 = 1, \; \hat{\gamma}_j = a_j \left(\prod_{i=2}^{j}(1 - a_i)\right)^{-1} \quad (j = 2, \ldots, n),$$

$$\hat{\beta}_n = \left(\prod_{i=2}^{n}(1 - a_i)\right)^{-1} = \hat{\gamma}_1 + \ldots + \hat{\gamma}_n \quad (n \in \mathbb{N});$$

compare Lemma 1.1. An important special case concerns $a_n = \frac{1}{n}$ with $\hat{\gamma}_n = 1$, $\hat{\beta}_n = n$ $(n \in \mathbb{N})$. This recursive variant of the Rosenblatt-Parzen method has been treated first by Wolverton and Wagner (1969) and by Yamato (1971).

4.3. Theorem. *Under the above assumptions and notations, let y be a continuity point of f.*
a) The sequence of estimates $f(\cdot, \ldots, \cdot, y)$ of $f(y)$ is asymptotically unbiased, i.e.

$$E\hat{f}_n(Y_1, \ldots, Y_n, y) \longrightarrow f(y) \quad (n \to \infty).$$

b) If $\sum a_n^2 h_n^{-1} < \infty$ – in the case $a_n = 1/n$ fulfilled for $h_n = n^{-\gamma}$ with $0 < \gamma < 1$ $(n \in \mathbb{N})$ –, then

$$\hat{f}_n(Y_1, \ldots, Y_n, y) \longrightarrow f(y) \quad (n \to \infty) \; a.s.$$

PROOF: a) By the above non-recursive representation one has

$$E\hat{f}_n(Y_1, \ldots, Y_n, y) = \frac{1}{\hat{\beta}_n} \sum_{j=1}^n \hat{\gamma}_j \frac{1}{h_j} \int_{\mathbb{R}} K\left(\frac{y-t}{h_j}\right) f(t) dt.$$

From this and

$$(5) \quad \frac{1}{h_n} \int_{\mathbb{R}} K\left(\frac{y-t}{h_n}\right) f(t) dt \longrightarrow f(y) \quad (n \to \infty),$$

which follows from Lemma 4.2 with $K^* = K$, the assertion is obtained.
b) One can use Theorem 4.1 with $k = 1$, $\vartheta = f(y)$,

$$X_n := \hat{f}_{n-1}(Y_1, \ldots, Y_{n-1}, y), \quad Z_n := \frac{1}{h_n} K\left(\frac{y-Y_n}{h_n}\right), \quad n \geq 2.$$

There (3) is verified by (5). (4) is verified by independence of (Y_n) and

$$h_n E Z_n^2 = \frac{1}{h_n} \int_{\mathbb{R}} K\left(\frac{y-t}{h_n}\right)^2 f(t) dt \longrightarrow f(y) \int_{\mathbb{R}} K(u)^2 du < \infty,$$

which is obtained by Lemma 4.2 with $K^* = K^2$. $\qquad \square$

References to refinements of the above results are e.g. Deheuvels (1974), Devroye (1979), see also Wertz (1985). – A recursive estimation of regression functions considered as elements of a Hilbert space by use of kernel functions has been treated by Révész (1973, 1977). As to recursive estimation of densities and regression functions, see also Prakasa Rao (1983).

The so-called passive stochastic approximation concerns recursive estimation of maximal points (modes) of density functions and extremal or zero points of regression functions in the case that the noise-contaminated function values can be only observed at random points whose choice cannot be controlled. Problems of this kind were treated e.g. by Fritz (1973), Härdle and Nixdorf (1987), Nazin, Polyak and Tsybakov (1989).

§ 5   Invariance principles in stochastic approximation

At first some definitions and results in the theory of distributional convergence in metric spaces are recalled; see the monographs of Billingsley (1968, 1971), Parthasarathy (1967), Gänssler and Stute (1977) with further references.

5.1. Definition. A complete separable metric space is called a Polish space (e.g. $\mathbb{R}^k$ with Euclidean norm, $C[0,1]$ with maximum norm). The minimal σ-algebra in a Polish space R which contains all open subsets of R, is the so-called Borel σ-algebra in R.

5.2. Definition. Let R be a Polish space with Borel σ-algebra and let X_n ($n \in \mathbb{N}$), X be R-valued random variables. (X_n) is said to converge to X in distribution – $X_n \xrightarrow{\mathcal{D}} X$ $(n \to \infty)$ – , if for each bounded continuous $g : R \to \mathbb{R}$

$$Eg(X_n \longrightarrow Eg(X)$$

or – equivalent – if for each [bounded] [P_X-a.e.] continuous $h : (R, \gamma) \longrightarrow (\mathbb{R}, \mathcal{B})$

$$h(X_n) \xrightarrow{\mathcal{D}} h(X)$$

holds.

5.3. Theorem (Skorokhod). *Let R be a Polish space with Borel σ-algebra and let X_n ($n \in \mathbb{N}$), X be R-valued random variables with*

$$X_n \xrightarrow{\mathcal{D}} X.$$

Then a probability space $(\Omega^*, \mathcal{A}^*, P^*)$ and R-valued random variables X_n^* ($n \in \mathbb{N}$), X^* on this probability space exist such that X_n^* and X_n have the same distribution ($n \in \mathbb{N}$), also X^* and X, and that

$$X_n^* \longrightarrow X^* \ a.s.$$

5.4. Lemma (Slutsky). *Let R be a real separable Banach space with Borel σ-algebra.*
a) If X_n, Y_n ($n \in \mathbb{N}$), X are R-valued random variables on a probability space and $c \in R$ with

$$X_n \xrightarrow{\mathcal{D}} X, \ Y_n \xrightarrow[P]{} c \ (i.e. \ in \ probability),$$

then

$$X_n + Y_n \xrightarrow{\mathcal{D}} X + c.$$

b) If X_n ($n \in \mathbb{N}$), X are R-valued random variables, Y_n ($n \in \mathbb{N}$) real-valued random variables on a probability space and $c \in \mathbb{R}$ with

$$X_n \xrightarrow{\mathcal{D}} X, \ Y_n \xrightarrow[P]{} c,$$

then

$$Y_n X_n \xrightarrow{\mathcal{D}} cX.$$

5.5. Definition. As standard Brownian motion or Wiener process (with state space $\mathbb{R}$) a $C[0,1]$-valued random variable $W = \{W_t; \ t \in [0,1]\}$ with the following properties is defined:

$$W_0 = 0, \qquad \underset{t\in(0,1]}{\forall} \quad W_t \quad N(0,t)\text{-distributed},$$

$$\underset{k\in\mathbb{N}}{\forall} \quad \underset{0\leq t_0 < t_1 < \dots < t_k \leq 1}{\forall} \quad (W_{t_1} - W_{t_0}, \dots, W_{t_k} - W_{t_{k-1}}) \text{ independent.}$$

An analogous definition concerns the case of an index set $\mathbb{R}_+$ via index subsets $[0,n]$, $n \in \mathbb{N}$.

The following theorem of Donsker type formulated for the martingale case states especially that the random course of a sequence of partial sums for i.i.d. square integrable real random variables centered at expectation (with piecewise linear interpolation) approximates the Wiener process after suitable standardizing (contraction of index or time axis at the ratio $1 : n$ and of the state axis at the ratio $1 : \sqrt{n}$). It is a so-called invariance principle or functional central limit theorem and contains a classical central limit theorem via an evaluation functional.

5.6. Theorem. *Let $(V_n)_{n\in\mathbb{N}}$ be a sequence of square integrable real random variables on a probability space $(\Omega, \mathcal{A}, P)$, $(\mathcal{A}_n)_{n\in\mathbb{N}_0}$ a nondecreasing sequence of sub-σ-algebras of $\mathcal{A}$ such that V_n is $\mathcal{A}_n$-$\mathcal{B}$-measurable $(n \in \mathbb{N})$. Further assume*

$$\underset{n\in\mathbb{N}}{\forall} \quad E(V_n|\mathcal{A}_{n-1}) = 0,$$

$$\frac{1}{n}\sum_{i=1}^{n} E(V_i^2|\mathcal{A}_{i-1}) \xrightarrow[P]{} 1,$$

$$\underset{\varepsilon>0}{\forall} \frac{1}{n}\sum_{i=1}^{n} E(V_i^2 \chi_{[|V_i|>\varepsilon\sqrt{n}]}|\mathcal{A}_{i-1}) \xrightarrow[P]{} 0 \quad (n \to \infty).$$

Let the $C[0,1]$-valued random variables Y_n, $n \in \mathbb{N}$, be defined by

$$Y_n(t) := \frac{1}{\sqrt{n}}\left[\sum_{i=1}^{[nt]} V_i + (nt - [nt])V_{[nt]+1}\right], \quad t \in [0,1].$$

Then

$$Y_n \xrightarrow{\mathcal{D}} W.$$

5.7. Definition. Let $\mathbb{B}$ be a real separable Banach space. A so-called Brownian motion in $\mathbb{B}$ or Wiener process with state space $\mathbb{B}$ and index space $[0,1]$ is a random variable $W = \{W_t; \ t \in [0,1]\}$ (also notation $\{W(t); \ t \in [0,1]\}$) with values in $C_{\mathbb{B}}[0,1]$ (with max-norm) and the property that for each bounded linear real-valued functional l on $\mathbb{B}$, $\{l\,W_t; \ t \in [0,1]\}$ is a Brownian motion differing from standard Brownian motion by the factor $[E(l\,W_1)^2]^{1/2}$.

An analogous definition concerns the case of an index set $\mathbb{R}_+$ via index subsets $[0, n]$, $n \in \mathbb{N}$.

The distributional invariance principle and the a.s. invariance principle given by Theorem 5.8 and Theorem 5.13, resp., and corresponding remarks concern the recursive scheme of Fabian (1968) by which various stochastic approximation procedures are comprehended in view of their convergence behaviour. The assumptions are especially analogous invariance principles for partial sum sequences, for which an example in the one-dimensional case is given by Theorem 5.6. Application of an evaluation functional in the assertion yields rate of convergence of the recursive sequence. References for invariance principles in stochastic approximation and for their assumptions and applications are Berger (1986), where the case of strong approximation is stressed, Walk (1977, 1988) and the literature cited in these articles, e.g. McLeish (1976), Kersting (1977), Lai and Robbins (1978), Ruppert (1982). See also references in §6.

5.8. Theorem. *Let $\mathbb{B}$ be a real separable Banach space and U_n, V_n, T $\mathbb{B}$-valued random variables, A_n $L(\mathbb{B})$-valued random variables $(n \in \mathbb{N})$ on a probability space $(\Omega, \mathcal{A}, P)$. Let $\beta \geq 0$ and $A \in L(\mathbb{B})$ such that*

(1) $\sigma^* := \min\{ \text{re } \lambda; \ \lambda \in \text{spec } (A)\} > \frac{\beta}{2}.$

Assume

(2) $U_{n+1} = U_n - \frac{1}{n} A_n U_n + n^{-\frac{1+\beta}{2}} V_n + n^{-(1+\frac{\beta}{2})} T, \quad n \in \mathbb{N},$

(3) $\|A_n - A\| \longrightarrow 0$ *a.s.* ,

further, with $\sigma := \min\{\sigma^*, \frac{1+\beta}{2}\}$,

(4) $\underset{\tau \in (0, \sigma - \frac{\beta}{2})}{\exists} \quad n^{-\tau} \cdot \sum_{k=1}^{n} k^{\tau - \frac{3}{2}} \|V_1 + \ldots + V_{k-1}\| = O_P(1),$

i.e. $\lim_{\tau \to \infty} \overline{\lim_{n}} \ P\left[n^{-\tau} \sum_{k=1}^{n} k^{\tau - \frac{3}{2}} \|V_1 + \ldots + V_{k-1}\| \geq r \right] = 0.$

For the $C_{\mathbb{B}}[0, 1]$-valued random variables Y_n $(n \in \mathbb{N})$ with

$$Y_n(t) := \frac{1}{\sqrt{n}} \left[\sum_{k=1}^{[nt]} V_k + (nt - [nt]) V_{[nt]+1} \right], \ t \in [0, 1],$$

it is assumed

(5) $Y_n \overset{\mathcal{D}}{\longrightarrow} W$ *(Brownian motion in $\mathbb{B}$).*

Let the Gaussian Markov process G with state space $\mathbb{B}$ be defined by

$G(t) := W(t) - (A - \frac{1+\beta}{2}) \int_{(0,1]} s^{A - \frac{\beta+3}{2}} W(st) ds, \ t \in [0, 1].$

For the $C_{\mathbb{B}}[0, 1]$-valued random variables Z_n, $n \in \mathbb{N}$, with

$Z_n(t) := \frac{1}{\sqrt{n}} \left[R_{[nt]} + (nt - [nt])(R_{[nt]+1} - R_{[nt]}) \right], \ t \in [0, 1],$

where

$R_n := \sqrt{n}(n^{\frac{\beta}{2}} U_{n+1} - (A - \frac{\beta}{2})^{-1} T), \ n \in \mathbb{N},$

it holds

$$Z_n \xrightarrow{\mathcal{D}} G,$$

especially

$$n^{\frac{\beta}{2}} U_{n+1} - (A - \tfrac{\beta}{2})^{-1} T \xrightarrow{\mathcal{D}} G(1).$$

5.9. Remark on Theorem 5.8. a) (4) is fulfilled under the other assumptions, if $E\|V_1 + \ldots + V_n\| = O(\sqrt{n})$ or if

$$\underset{\tau' \in (0, \sigma - \frac{\beta}{2})}{\exists} \quad n^{-\tau'} \underset{k=1,\ldots,n}{\max} \; k^{\tau' - \frac{1}{2}} \|V_1 + \ldots + V_k\| = O_p(1).$$

The latter follows from a sufficient condition for $(4) \wedge (5)$ given by Berger (1986, p. 529).

b) The spectral condition (1) guarantees the existence of G. Other representations of G are given by

$$G(t) = \int\limits_{(0,1]} s^{A - \frac{\beta+1}{2}} \, dW(ts), \quad t \in [0,1],$$

and

$$G(0) = 0, \; G(t) = t^{-A + \frac{\beta+1}{2}} \int\limits_{(0,t]} v^{A - \frac{\beta+1}{2}} \, dW(v), \quad t \in (0,1].$$

G satisfies the stochastic differential equation

$$dG(t) = (-A + \tfrac{\beta+1}{2}) \tfrac{1}{t} G(t) dt + dW(t), \quad t \in (0,1].$$

c) In the recursion formula (2), T may be replaced by $\mathbb{B}$-valued random variables T_n with $T_n \to T$ a.s. This is apparent, because $n^{-1/2}(T_n - T)$ may be added to V_n.

d) In the case that $\mathbb{B}$ is a Hilbert space, $G(1)$ is a $\mathbb{B}$-valued Gaussian random variable with expectation vector zero and a covariance operator Cov $G(1)$ satisfying

$$(A - \tfrac{\beta}{2}) \; \text{Cov} \; G(1) + \; \text{Cov} \; G(1)(A' - \tfrac{\beta}{2}) = \; \text{Cov} \; W(1).$$

5.10. Remark. Theorem 5.8 also holds if (3) together with (4) is replaced by

$$\|\tfrac{1}{n}(A_1 + \ldots + A_n) - A\| \longrightarrow 0 \text{ a.s. },$$

$$\tfrac{1}{n}(\|A_1\| + \ldots + \|A_n\|) = O(1) \text{ a.s. },$$

$$\underset{\tau' \in (0, \sigma - \frac{\beta}{2})}{\exists} \quad n^{-\tau'} \underset{k=1,\ldots,n}{\max} \; k^{\tau' - \frac{1}{2}} \|V_1 + \ldots + V_k\| = O_P(1).$$

(Walk 1988). Thus, with $\beta = 1$ and $T_n = T = 0$, for the stochastic recursion of Theorem 2.1 with gains $a_n = 1/n$, a result on rate of convergence is established by setting $V_n = b_n - A_n \vartheta$ there. — Let under these assumptions, with $\sigma^* > 0$ (only), $0 < \tau' < 1/2$ and $(b_1 + \ldots + b_n)/n \to b$ a.s., a stochastic recursion be defined by

$$X_{n+1} = X_n - \frac{1}{n} B_n (A_n X_n - b_n)$$

in $\mathbb{B}$ with $L(\mathbb{B})$-valued random variables B_n satisfying the auxiliary recursion

$$B_{n+1} = B_n - \frac{c}{n}(A_n B_n - 1), \quad 0 < c < \infty.$$

Then, by use of Theorem 2.1 and the first part of this Remark together with partial summation, the auxiliary recursion and Lemma 3 in Walk (1988) — there with $A_k B_k - 1$ instead of $A_k - A$ — , one obtains

$$\|B_n - A^{-1}\| \to 0 \text{ a.s.,}$$

$$X_n \to \vartheta \text{ a.s.,}$$

$$\sqrt{n}(X_n - \vartheta) \xrightarrow{D} A^{-1} W(1).$$

In the Hilbert space case the procedure can be considered as a modification of the (Newton-type) recursive least squares method with the same optimal asymptotic covariance operator A^{-1} Cov $W(1)$ $(A^{-1})'$ (minimal trace); as to the latter method see e. g. Ljung and Söderström (1983) and part III of this monograph.

5.11. Remark. Let in Theorem 5.8 and Remarks 5.9, 5.10 especially $\mathbb{B} = \mathbb{R}$ and thus $A \in \mathbb{R}$, and let V_n in the recursion formula be replaced by δV_n with $\delta \in (0, \infty)$ and W be the standard Brownian motion with index space $[0, 1]$ or $\mathbb{R}_+$. Then in the representation of G, trivially W has to be endowed with the factor δ. Instead of this G, as limit random variable in distributional convergence also the Gaussian process G^* with

$$G^*(0) = 0, \ G^*(t) = t^{A - \frac{\beta-1}{2}} W(\tfrac{1}{2A-\beta}\delta^2 t^{-2A+\beta}), \quad t \in (0, 1]$$

can appear (for $\beta = 1$ see McLeish (1976)), which is stochastically equivalent to G. The equivalence is shown by Remark 5.9b via covariance functions.

In the proof of Theorem 5.8 the following simple lemma will be used.

5.12. Lemma. *If for nonnegative-real random variables* $L_n, Y_{n,\delta}, Z_{n,\delta}$ *with* $L_n \leq Y_{n,\delta} + Z_{n,\delta}$ $(n \in \mathbb{N}, \ \delta > 0)$
the relations

$$\underset{\varepsilon > 0}{\forall} \quad \underset{\delta \to 0}{\overline{\lim}} \ \ \overline{\lim_{n}} \ \ P[Y_{n,\delta} \geq \varepsilon] = 0,$$

$$\underset{\delta > 0}{\forall} \quad Z_{n,\delta} \xrightarrow{P} 0 \quad (n \to \infty)$$

hold, then

$$L_n \xrightarrow{P} 0 \quad (n \to \infty).$$

PROOF OF THEOREM 5.8: In the first step the case $A_n = A \ (n \in \mathbb{N})$, $\beta = 1$, $T = 0$, $U_1 = 0$ is treated. The recursion yields

$$U_{n+1} = \sum_{k=1}^{n} \left[\tfrac{1}{k} \prod_{j=k+1}^{n} (1 - \tfrac{A}{j}) \right] V_k$$

and then

$$(6) \quad U_{n+1} = \tfrac{1}{n} S_n + (1 - A) \sum_{k=1}^{n-1} \tfrac{1}{k(k+1)} \prod_{j=k+2}^{n} (1 - \tfrac{A}{j}) S_k,$$

where

$$S_k := \sum_{j=1}^{k} V_j \quad (k \in \mathbb{N}).$$

With
$$Z_n^*(t) := \tfrac{1}{\sqrt{n}} R_{[nt]}, \quad Y_n^*(t) := \tfrac{1}{\sqrt{n}} S_{[nt]}, \quad t \in [0,1],$$
one now obtains

(7) $Z_n^*(t) = Y_n^*(t) + (1-A)[nt] \sum_{k=1}^{[nt]-1} \frac{1}{k(k+1)} \prod_{j=k+2}^{[nt]} (1-\frac{A}{j}) Y_n^*(\frac{k}{n}).$

Because of Theorem 5.3 without loss of generality $Y_n \to W$ a.s. and thus

(8) $\sup_{t \in [0,1]} \|Y_n^*(t) - W(t)\| \longrightarrow 0$ a.s.

may be assumed. (4) yields for some $\tau \in (0, \sigma - \frac{1}{2})$

(9) $\underset{\varepsilon > 0}{\forall} \ \underset{\delta \to 0}{\overline{\lim}} \ \overline{\underset{n}{\lim}} \ P\left[n^{-\tau + \frac{1}{2}} \sum_{k=1}^{[n\delta]} k^{\tau - \frac{3}{2}} \|Y_n(\frac{k}{n})\| \geq \varepsilon \right] = 0.$

Now

(10) $\sup_{t \in [0,1]} \|Z_n^*(t) - G(t)\| \underset{P}{\longrightarrow} 0$

and thus $Z_n \overset{D}{\longrightarrow} G$ will be derived. Because of (7) and (8), it suffices to show

(11) $\sup_{t \in (0,1]} \| \sum_{k=1}^{[nt]-1} F_{nk}(t) - \int_{(0,t]} F(v,t) dv \| \underset{P}{\longrightarrow} 0,$

where

$F_{nk}(t) = [nt] \frac{1}{k(k+1)} (1-A) \prod_{j=k+2}^{[nt]} (1 - \frac{A}{j}) Y_n^*(\frac{k}{n})$

for $k \in \{1, \dots, [nt]-1\}$, $n \in \mathbb{N}$, $t \in (0,1]$,

$F(v,t) = (1-A) t^{1-A} v^{A-2} W(v)$ for $0 < v \leq t$, $t \in (0,1]$.

The left side of (11) is dominated by $Y_{n,\delta} + Z_{n,\delta}$ with arbitrary $\delta \in (0,1]$, where

$Y_{n,\delta} := \sup_{t \in (0,\delta]} \sum_{k=1}^{[nt]-1} \|F_{nk}(t)\| + \sup_{t \in (0,\delta]} \int_{(0,t]} \|F(v,t)\| dv$

$+ \sup_{t \in (\delta,1]} \sum_{k=1}^{[n\delta]-1} \|F_{nk}(t)\| + \sup_{t \in (\delta,1]} \int_{(0,\delta]} \|F(v,t)\| dv,$

$Z_{n,\delta} := \sup_{t \in (\delta,1]} \| \sum_{k=[n\delta]}^{[nt]-1} F_{nk}(t) - \int_{(\delta,t]} F(v,t) dv \|.$

For $Y_{n,\delta}$ and $Z_{n,\delta}$ the assumptions of Lemma 5.12 can be verified. In fact, (1) yields that for each $\mu \in (0, \sigma^*)$ a $c \in \mathbb{R}_+$ exists such that

$\| \prod_{j=k}^{n} (1 - \frac{A}{j}) \| \leq c(\frac{k}{n})^\mu$ for all $k, n \in \mathbb{N}$ with $k \leq n$

and

$\|(\frac{v}{t})^A\| \leq c(\frac{v}{t})^\mu$ for all $v, t > 0$ with $v \leq t$

(Daleckii and Krein 1970/1974, section I. 4; Walk and Zsidó 1989; see also §1). One chooses $\mu = \tau + \frac{1}{2}$, then obtains

$$Y_{n,\delta} \leq c^* \left[n^{-\tau+\frac{1}{2}} \sum_{k=1}^{[n\delta]-1} k^{\tau-\frac{3}{2}} \|Y_n(\tfrac{k}{n})\| + \int_{(0,\delta]} v^{\tau-\frac{3}{2}} \|W(v)\| dv \right]$$

for some $c^* \in \mathbb{R}_+$ and applies (9). Finally a pathwise consideration shows $Z_{n,\delta} \to 0$ $(n \to \infty)$ a.s. for each $\delta \in (0,1]$, where the relation

$$\underset{\delta \in (0,1]}{\forall} \quad \sup_{\delta \leq v \leq t \leq 1} \quad \left\| \prod_{j=[nv]+2}^{[nt]} (1 - \tfrac{A}{j}) - (\tfrac{v}{t})^A \right\| \to 0 \; (n \to \infty)$$

and (8) are used.

In the second step the more general case $A_n \to A$ a.s., but with $\beta = 1$, $T = 0$, $U_1 = 0$, is treated. If the random variables Z_n in the special case $A_n = A$ $(n \in \mathbb{N})$ are denoted by Z_n^0, then according to Lemma 5.4a the problem is reduced to show $Z_n - Z_n^0 \xrightarrow[P]{} 0$, because of $Z_n^0 \xrightarrow{D} G$ (see first step). One uses (6) and its generalization and argues similarly to the first step, where without loss of generality $A_n \to A$ uniformly may be assumed. Analogously, but more easily, the case of general U_1 can be treated.

In the last step one reduces the general case to the case $\beta = 1$, $T = 0$ by use of a recursion for the random variables

$$U'_{n+1} := n^{\frac{-1+\beta}{2}} U'_{n+1} - n^{-\frac{1}{2}} (A - \tfrac{\beta}{2})^{-1} T \quad (n \in \mathbb{N}). \qquad \square$$

The following almost sure invariance principle (Mark 1982) corresponds to the distributional invariance principle of Theorem 5.8.

5.13. Theorem. *Let $\mathbb{B}$ be a real separable Banach space, $A \in L(\mathbb{B})$ with (1), $\beta \geq 0$. Let further on a probability space $(\Omega, \mathcal{A}, P)$ $\mathbb{B}$-valued random variables U_n, V_n, T and $L(\mathbb{B})$-valued random variables A_n $(n \in \mathbb{N})$, a Brownian motion $W = \{W(t); \; t \in \mathbb{R}_+\}$ with state space $\mathbb{B}$ and, analogously to Theorem 5.8, a Gaussian Markov process $G = \{G(t); \; t \in \mathbb{R}_+\}$ be defined. Assume (2), (3) and*

$$(12) \quad (n \; loglog \; n)^{-\frac{1}{2}} \|\sum_{k=1}^{n} V_k - W(n)\| \to 0 \; (n \to \infty) \; a.s.$$

Then

$$(t \; loglog \; t)^{-\frac{1}{2}} \| [t]^{\frac{1+\beta}{2}} U_{[t]+1} - G(t)\| \to 0 \; (t \to \infty) \; a.s.$$

5.14. Remark on Theorem 5.13. a) In the recursion formula (2), T may be replaced by T_n with

$$n^{-1}(loglog \; n)^{-\frac{1}{2}} \|\sum_{k=1}^{n} T_k\| \to 0 \; (n \to \infty) \; \text{a.s.}$$

b) (3) may be weakened to the ergodicity assumptions

$$\|\tfrac{1}{n}(A_1 + \ldots + A_n) - A\| \to 0 \; \text{a.s.} \; ,$$

$$\tfrac{1}{n}(\|A_1\| + \ldots + \|A_n\|) = O(1) \; \text{a.s.} \; ,$$

c) If (12) is weakened to

$$(n \; loglog \; n)^{-\frac{1}{2}} \|\sum_{k=1}^{n} (V_k + k^{-1/2} T_k)\| = O(1) \; \text{a.s.} \; ,$$

then the assertion is weakened to

$$(\log\log n)^{-\frac{1}{2}} n^{\frac{\beta}{2}} \|X_n\| = O(1) \text{ a.s.}$$

In the following several applications of Theorem 5.8 to stochastic approximation processes shall be given, for reasons of simplicity mostly in the one-dimensional case. Literature concerning the multi-dimensional case is cited in the above mentioned articles.

The first application concerns the Robbins-Monro process. The assumption that the sequence of estimates a.s. converges to a zero point of the regression function has been treated in §1. As to a verification of the assumptions on the measurement errors, it is referred to Theorem 5.6 and Remark 5.9a.

5.15. Theorem. *Let* $f : \mathbb{R} \to \mathbb{R}$ *be measurable with zero point* $\vartheta \in \mathbb{R}$ *and* X_n, W_n $(n \in \mathbb{N})$ *be real random variables with*

$$X_{n+1} = X_n - \frac{c}{n} f(X_n) + \frac{c}{n} W_n,$$

where $c \in (0, \infty)$. *Assume* f *differentiable at* ϑ *with* $f'(\vartheta) > \frac{1}{2c}$ *and* $X_n \to \vartheta$ *a.s. Further assume that* $V_n := \frac{1}{\rho} W_n$ *with suitable* $\rho \in (0, \infty)$ *fulfills the corresponding assumptions of Theorem 5.8. Then for* $U_n := X_n - \vartheta$ *the assertion of Theorem 5.8 holds in the version given by Remark 5.11, with* $A = cf'(\vartheta)$, $\beta = 1$, $T = 0$, $\delta = c\rho$; *especially*

$$\sqrt{n}(X_n - \vartheta) \xrightarrow{\mathcal{D}} N(0, \tfrac{c^2\rho^2}{2cf'(\vartheta)-1}) \text{ -distributed random variable.}$$

PROOF: The differentiability assumption together with $X_n \to \vartheta$ a.s. yields

$$X_{n+1} - \vartheta = X_n - \vartheta - \frac{c}{n}(f'(\vartheta) + o(1))(X_n - \vartheta) + \frac{c}{n} W_n \text{ a.s.}$$

Now an immediate application of Theorem 5.8, Remark 5.9d and Remark 5.11 is possible.   □

5.16. Remark. a) In view of minimal variance in the limit distribution, the asymptotically optimal c in Theorem 5.15 is $1/f'(\vartheta)$. The variance is then $\rho^2/f'(\vartheta)^2$. In the so-called adaptive procedures of stochastic approximation, e. g. Venter method and Anbar-Lai-Robbins method, which are Newton-type stochastic algorithms, the optimal c is estimated in the course of iterations by use of divided differences or of the slope of the least squares line with truncation (Venter 1967, Fabian 1968 and 1973, Ruppert 1982, Schwabe 1986; Anbar 1978, Lai and Robbins 1981). For the m-dimensional case, where $(Df(\vartheta))^{-1}$ plays the role of $1/f'(\vartheta)$, see Wei (1987) with further references; compare also Remark 5.10. In view of optimality see also Theorem 5.21 below.
b) If in Theorem 5.15 the observation errors $-W_n$ conditioned with respect to the past possess a unitary differentiable density g with Fisher information $I(g) := \int (g'/g)^2 g \in (0, \infty)$ and if a transform $h(f(X_n) - W_n)$ is used instead of $c(f(X_n) - W_n)$, then under suitable assumptions an h optimal with respect to the variance of the limit normal distribution is given as $-g'/g$, where the minimal variance is $I(g)^{-1} f'(\vartheta)^{-2}$ (Abdelhamid 1971, Anbar 1971). By this, in the case of a linear f an asymptotically efficient estimate of the parameter ϑ is

given. Fabian 1973 showed that a corresponding result can be obtained without prior knowledge of g which is assumed to be symmetric; in 1983 he showed a sharper optimality property on the basis of the asymptotic estimation theory of LeCam and Hájek.

Now an application to the Kiefer-Wolfowitz process will be given.

5.17. Theorem. *Let $f : \mathbb{R} \to \mathbb{R}$ be differentiable with $\vartheta \in \mathbb{R}$ as zero point of f' and X_n, W_n ($n \in \mathbb{N}$) be real random variables with*

$$X_{n+1} = X_n - \frac{c}{n} \frac{f(X_n + c'n^{-\gamma}) - f(X_n - c'n^{-\gamma}) - 2W_n}{2c'n^{-\gamma}},$$

where $c \in (0, \infty)$, $c' \in (0, \infty)$, $\gamma = \frac{1}{4}$. Assume f two times differentiable at ϑ with $f''(\vartheta) > \frac{1}{4c}$ and $X_n \to \vartheta$ a.s. Further assume that $V_n := \frac{1}{\rho}W_n$ with suitable $\rho \in (0, \infty)$ fulfills the corresponding assumptions of Theorem 5.8. Then for $U_n := X_n - \vartheta$ the assertion of Theorem 5.8 holds in the version given by Remark 5.11, with $A = cf''(\vartheta)$, $\beta = \frac{1}{2}$, $T = 0$, $\delta = \frac{c\rho}{c'}$; especially

$$n^{\frac{1}{4}}(X_n - \vartheta) \xrightarrow{\mathcal{D}} N(0, \frac{c^2\rho^2}{c'^2(2cf''(\vartheta) - \frac{1}{2})}) \text{ -distributed random variable.}$$

PROOF: One notices

$$f(\vartheta + h + q) - f(\vartheta + h - q)$$

$$= \int_{-1}^{1} f'(\vartheta + h + tq)dtq$$

$$= 2f''(\vartheta)hq + o(1)hq + o(1)q^2, \ (h, q) \to (0, 0),$$

valid because of the differentiability assumptions, employs $X_n \to \vartheta$ a.s. and obtains the assertion by Theorem 5.8 together with Remark 5.9c,d and Remark 5.11. $\qquad\square$

5.18. Remark. If in Theorem 5.17 additionally f is assumed three times differentiable at ϑ, but with $f''(\vartheta) > \frac{1}{3c}$, and if in the recursion formula $\gamma = \frac{1}{6}$ is chosen, then for $U_n := X_n - \vartheta$ the assertion of Theorem 5.8 holds in the version given by Remark 5.11, with $A = cf''(\vartheta)$, $\beta = \frac{2}{3}$, $T = -\frac{c}{6}c'^2f'''(\vartheta)$, $\delta = \frac{c\rho}{c'}$; especially

$$n^{\frac{1}{3}}(X_n - \vartheta) \xrightarrow{\mathcal{D}} N\left(\frac{-cc'^2f'''(\vartheta)}{6cf''(\vartheta) - 2}, \frac{c^2\rho^2}{c'^2(2cf''(\vartheta) - \frac{2}{3})}\right) \text{ -distributed vari-}$$

able.

This is obtained analogously to the proof of Theorem 5.17 by an expansion of f around ϑ.

5.19. Remark. The choice of γ in Theorem 5.17 and Remark 5.18, resp., is optimal in view of convergence order. A corresponding order of convergence, under a modified set of assumptions, is valid for convergence in L_2 mean (see e.g. Schmetterer 1969, Herkenrath 1980). – As to the conflict between the goal of estimating a zero point or extremum point ϑ of a regression function well by X_n and the goal of low control costs $\sum_{i=1}^{n}(X_i - \vartheta)^2$ see e.g. the survey article of Ruppert (1991).

The above invariance principle for the Robbins-Monro process (Theorem 5.15) leads in the following to an asymptotic fixed width confidence interval, in the sense of Chow and Robbins, for a zero point of a regression function (McLeish 1976).

5.20. Theorem. *Let the assumptions of Theorem 5.15 hold.*
a) For a sequence (N_n) of $\mathbb{N}$-valued random variables and a sequence (b_n) of positive real numbers with $b_n \to \infty$ and $N_n/b_n \xrightarrow[P]{} 1$ $(n \to \infty)$, it holds

$$N_n^{1/2}(X_{N_n} - \vartheta) \xrightarrow{\mathcal{D}} N(0, \tfrac{c^2\rho^2}{2cf'(\vartheta)-1}) \text{ -distributed random variable.}$$

b) Let (R_n^2) be a strongly consistent estimation sequence for

$$R^2 := \tfrac{c^2\rho^2}{2cf'(\vartheta)-1}, \text{ i.e. } R_n^2 \to R^2 \text{ a.s.}$$

(in course of the iterations obtainable via strongly consistent estimation sequences for ρ^2 and $f'(\vartheta)$). Let $u_{\alpha/2}$ denote the $\alpha/2$ -fractile of $N(0,1)$ and let

$$N(d) := \inf\{n \in \mathbb{N}; \ R_n^2 + \tfrac{1}{n} \leq n(\tfrac{d}{u_{\alpha/2}})^2\}$$

(additional term $\tfrac{1}{n}$ for practical reasons),

$$b(d) := \inf\{n \in \mathbb{N}; \ R^2 \leq n(\tfrac{d}{u_{\alpha/2}})^2\}, \ d \in (0, \infty).$$

Then

$$\tfrac{N(d)}{b(d)} \to 1 \ (d \to 0) \text{ a.s.},$$

$$N(d)^{1/2} \tfrac{X_{N(d)} - \vartheta}{R} \xrightarrow{\mathcal{D}} N(0,1) \text{ -distributed random variable } (d \to 0),$$

$$P[[X_{N(d)} - d, \ X_{N(d)} + d] \ni \vartheta] \to 1 - \alpha \ (d \to 0),$$

i.e. with respect to $d \to 0$ an asymptotic confidence interval of fixed length $2d$ and of size $1 - \alpha$ for the unknown ϑ, where ρ^2 and $f'(\vartheta)$ are unknown, is given by a realization of $[X_{N(d)-d}, \ X_{N(d)+d}]$.

PROOF: Without loss of generality $N_n/b_n \to 1$ uniformly, with $1/2 \leq N_n/b_n \leq 3/2$, may be assumed. By Theorem 5.15, $Z_{2[b_n]} \xrightarrow{\mathcal{D}} G$. According to Billingsley (1968, Theorem 4.4), from this and $N_n/(2b_n) \xrightarrow[P]{} 1/2$, it follows that the sequence of $C[0,1] \times [1/4, 3/4]$ -valued random variables $(Z_{2[b_n]}, N_n/(2[b_n]))$ converges in distribution to $(G, 1/2)$. Using the continuous function $(x, y) \to y^{-1/2}x(y)$, $(x, y) \in C[0,1] \times [1/4, 3/4]$, and noticing that $2^{1/2}G(1/2)$ and $G(1)$ have the same distribution (according to Remark 5.11), one obtains the assertion in a) and then, by Lemma 5.4b, in b). $\qquad\square$

Another application of the invariance principle for the Robbins-Monro process concerns use of the integral functional. Under the assumptions of Theorem 5.15, for the sequence of arithmetic means of the estimates it yields

$$\sqrt{n}(\tfrac{X_1 + \ldots + X_n}{n} - \vartheta) \xrightarrow{\mathcal{D}} \int_0^1 v^{-1}G(v)dv,$$

where the limit random variable is $N(0, \frac{2c\rho^2}{(2cf'(\vartheta)-1)f'(\vartheta)})$ -distributed (Pechtl 1988). The variance of the limit distribution converges to the optimal value $\rho^2/f'(\vartheta)^2$ for $c \to \infty$.

Here a connection exists with the result of Ruppert (1988, 1991) that under regularity assumptions for a corresponding Robbins-Monro process (X'_n) with weight factor $n^{-\alpha}$ ($\frac{1}{2} < \alpha < 1$) instead of $1/n$ one has

$$\sqrt{n}\left(\frac{X'_1+...+X'_n}{n} - \vartheta\right) \xrightarrow{D} N(0, \frac{\rho^2}{f'(\vartheta)^2})$$ -distributed random variable.

This surprising result tells that taking arithmetic means on a Robbins-Monro process with convergence order less than $1/\sqrt{n}$ yields the optimal convergence order $1/\sqrt{n}$ with optimal asymptotic variance, without an estimation of $f'(\vartheta)^{-1}$ used in the adaptive procedures. A corresponding result, also for linear filtering and regression, has independently been obtained by Polyak (1990) in the m-dimensional case (minimal trace of the asymptotic covariance matrix), see also Polyak and Juditsky (1990) for a more general form containing the following theorem. Further investigations by several authors have been stimulated.

5.21. Theorem. *Let* $f : \mathbb{R}^m \to \mathbb{R}^m$ *be measurable with zero point* $\vartheta \in \mathbb{R}^m$, *and let* X_n, W_n *be* m-*dimensional random vectors and* a_n *be positive real numbers with*

$$X_{n+1} = X_n - a_n f(X_n) + a_n W_n, \quad n \in \mathbb{N}.$$

Assume

$$\text{spec}\,(A) \subset \{z \in \mathbb{C};\ re\,z > 0\},$$
$$\|f(x) - A(x-\vartheta)\| = O(\|x-\vartheta\|^{1+\lambda})$$

for $A := Df(\vartheta)$, *further square integrability of* W_n *and*

$$E(W_{n+1}|X_1, W_1, \ldots, W_n) = 0,$$
$$\sup_n E(\|W_{n+1}\|^2\chi_{[\|W_{n+1}\|>c]}|X_1, W_1, \ldots, W_n) \xrightarrow{P} 0 \quad (c \to \infty)$$
$$E(W_n W'_n|X_1, W_1, \ldots, W_n) \xrightarrow{P} S,$$

where S *is a positive-semidefinite* $m \times m$-*matrix. If*

$$\frac{a_n - a_{n+1}}{a_n} = o(a_n), \quad \sum a_n^{\frac{1+\lambda}{2}}/\sqrt{n} < \infty$$

and

$$X_n \to \vartheta \ a.s,$$

then

$$\sqrt{n}\left(\frac{X_1+...+X_n}{n} - \vartheta\right) \xrightarrow{D} N(0, A^{-1}S(A^{-1})')\text{-}distributed\ random\ vector.$$

In the linear case $f(x) \equiv A(x-\vartheta)$ *this assertion also holds, if* $a_n \equiv \alpha \in (0, 2\min\{re\,(1/\mu);\ \mu \in spec\,A\})$.

The following heuristic argument for the last part of Theorem 5.21 is formulated for the continuous case with white noise and immediately yields for the discrete case $(X_1 + \ldots + X_n)/n \to \vartheta$ a.s. if the random vectors W_n only satisfy $(W_1 + \ldots + W_n)/n \to 0$ a.s. Let X satisfy the stochastic differential equation

$$dX(t) = -\alpha A X(t) dt + \alpha dW(t), \quad t \geq 1,$$

in $\mathbb{R}^m$ with $X(1) = 0$, where W denotes Brownian motion. Then Y with

$$Y(t) := \frac{1}{t} \int_1^t X(s) ds, \quad t \geq 1,$$

satisfies

$$dY(t) = (-\alpha A - \tfrac{1}{t}) Y(t) dt + \tfrac{\alpha}{t} (W(t) - W(1)) dt.$$

The convergence results

$$Y(t) \quad \to \quad 0 \text{ a.s.},$$
$$\sqrt{t} Y(t) \quad \overset{\mathcal{D}}{\to} \quad N(0, A^{-1} \text{Cov} W(1) (A^{-1})')\text{-distributed random vector}$$

now follow from the corresponding results for Z with

$$Z(t) := \alpha e^{-\alpha A t} \int_1^t e^{\alpha A s} \frac{W(s)}{s} ds, \quad t \geq 1,$$

which satisfies

$$dZ(t) = -\alpha A Z(t) dt + \tfrac{\alpha}{t} W(t) dt.$$

5.22. Remark. Frees and Ruppert 1991 (see also Ruppert 1991) propose a least squares method following the Robbins-Monro procedure with deterministic or – in an adaptive variant - random gains of order $1/n$ for estimation of a zero point ϑ of a regression function f. If for the latter procedure, in the one-dimensional case, the usual conditions for asymptotic normality with convergence order $1/\sqrt{n}$ are fulfilled, then the zero point of the least squares regression line constructed for the Robbins-Monro process $X_1, \ldots, X_n$ with noise-corrupted observations $f(X_1) - W_1, \ldots, f(X_n) - W_n$, yields an estimate of ϑ for which the authors establish asymptotic normality with convergence order $1/\sqrt{n}$ and minimal asymptotic variance. Here the stability properties and also the order $\log n$ of control costs of the Robbins-Monro type procedure with gains of order $1/n$ and without additional observations (see Lai and Robbins 1981, compare Remark 5.19) are preserved. The proposed method in some sense plays an intermediate role between the Anbar-Lai-Robbins method and the Ruppert-Polyak-Juditsky method. For, in the case of gains c/n, especially with $c > 1/(2f'(\vartheta))$, the estimator of ϑ can be written in the form $c^{-1} B_n X_{n+1} + (1 - c^{-1} B_n) \overline{X}_n$, where $\overline{X}_n := (X_1 + \ldots + X_n)/n$ and B_n^{-1} is the slope of the regression line. – Now consider the situation of Theorem 5.15 in its immediate generalization (by Theorem 5.8) to a real separable Hilbert space $\mathbb{H}$, where $f : \mathbb{H} \to \mathbb{H}$ is assumed Fréchet differentiable at its zero point ϑ with

$$\min\{\text{re } \lambda; \ \lambda \in \text{spec } A\} > \frac{1}{2}, \quad A := cDf(\vartheta).$$

Let (B_n) be a strongly or weakly consistent sequence of estimates of $(Df(\vartheta))^{-1}$, possibly based on $X_1, \ldots, X_n$ and $f(X_1) - W_1, \ldots, f(X_n) - W_n$ and possibly defined in a recursive manner. Then the generalized Theorem 5.15 yields, by use of Theorems 4.4 and 5.1 in Billingsley (1968),

$$\sqrt{n}[c^{-1}B_n X_{n+1} + (1 - c^{-1}B_n)\overline{X}_n - \vartheta]$$

$$\xrightarrow{\mathcal{D}} A^{-1}G(1) + (1 - A^{-1})\int_0^1 v^{-1}G(v)dv = A^{-1}W(1),$$

the last relation following from Remark 5.9b (compare Mark 1982, p. 70). Moreover a distributional invariance principle with limit process $A^{-1}W$ is obtained.

Finally an application of the last assertion of Theorem 5.8 to rate of convergence in recursive density estimation will be given (Yamato 1971). The notations of §4 will be used. As there an approximation-theoretic lemma, now on the quality of approximation, is employed.

5.23. Lemma. *Let $f : \mathbb{R} \to \mathbb{R}_+$ be a bounded density that is two times differentiable at $y \in \mathbb{R}$. Let further $K : \mathbb{R} \to \mathbb{R}_+$ be an even density function with $\int u^2 K(u)du < \infty$. For the singular integrals*

$$g_v(y) := \frac{1}{v} \int_{\mathbb{R}} K(\frac{y-t}{v})f(t)dt \quad (y \in \mathbb{R}, v > 0),$$

it holds

$$\frac{1}{v^2}(g_v(y) - f(y)) \to \frac{1}{2}f''(y)\int_{\mathbb{R}} u^2 K(u)du \quad (v \to 0).$$

PROOF: Similarly to the proof of Lemma 4.2, by use of an expansion of f around ϑ. □

5.24. Theorem. *Let the assumptions of Theorem 4.3 hold, further assume f two times differentiable at ϑ. If K is an even density function with $M := \int u^2 K(u)du < \infty$ and if $a_n = 1/n$ and $h_n = c'n^{-1/5}$ with $c' > 0$ are chosen $(n \in \mathbb{N})$, then*

$$n^{2/5}(\hat{f}_n(Y_1, \ldots, Y_n, y) - f(y))$$

$$\xrightarrow{\mathcal{D}} N(\frac{5}{6}c'^2 f''(y)M, \frac{5}{6c'}f(y)M) \text{ -distributed random variable.}$$

PROOF: As in the proof of Theorem 4.3b one sets $\vartheta := f(y)$,
$X_n := \hat{f}_{n-1}(Y_1, \ldots, Y_{n-1}, y)$, $Z_n := \frac{1}{h_n}K(\frac{y-Y_n}{h_n})$,
further $U_n := X_n - \vartheta$, $\delta := (\frac{1}{c'}f(y)\int_{\mathbb{R}} K(u)^2 du)^{1/2}$,

$$V_n := \delta^{-1}n^{-\frac{1}{10}}(Z_n - EZ_n), \quad T_n := n^{\frac{2}{5}}(EZ_n - \vartheta)$$

and obtains the recursion

$$U_{n+1} = (1 - \frac{1}{n})U_n + n^{-\frac{9}{10}}\delta V_n + n^{-\frac{7}{5}}T_n, \quad n \geq 2.$$

Now Theorem 5.8 together with Remark 5.9a, c, d and Remark 5.11 is applicable with $A_n = A = 1$, $\beta = 4/5$, $T = c'^2 f''(y)M/2$. For the sequence (V_n) is independent, $|V_n| \leq$ const $n^{1/10}$, $EV_n^2 \to 1$ by Lemma 4.2, thus $E(V_1 + \ldots + V_n)^2/n \to 1$, and condition (5) of Theorem 5.8 (there Y_n in another meaning) is fulfilled by Theorem 5.6; further $T_n \to T$ by Lemma 5.23. □

It should be noticed that in the above theorems the convergence order of the Robbins-Monro process with differentiability of the regression function f is $n^{-1/2}$, the convergence order of the Kiefer-Wolfowitz process with two (three) times differentiability of f is $n^{-1/4}$ $(n^{-1/3})$, the convergence order of recursive density estimation with two times differentiability of the density function is $n^{-2/5}$.

§ 6   On the theory of large deviations

Concerning the Robbins-Monro process (X_n) for estimation of a zero point ϑ of a regression function, Woodroofe (1972) in the one-dimensional case and Révész (1973), Schmetterer (1976) in the Hilbert space case established exponential bounds for probabilities $P[\|X_n - \vartheta\| \geq \varepsilon]$ $(\varepsilon > 0)$ of large deviations.

A functional theory of large deviations for Robbins-Monro and Kiefer-Wolfowitz processes has been developed by Kushner (1984), Dupuis and Kushner (1985, 1987). It regards the remainder of the paths of the process (X_n) from large indices on and the shape of the regression function, not only locally as in the functional central limit theorems treated in §5. In fact, the connection between the stochastic iteration $X_{n+1} = X_n - a_n f(X_n) + a_n V_n$ with small noises $a_n V_n$, the differential equation $\dot{x} = -f(x)$ and related differential equations with small random noise allows not only an investigation of the above recursion in view of a.s. convergence (see §1) and of functional central limit theorems (see for a rather general situation Kushner and Huang 1979 and Bouton 1988; compare also Remark 5.9b), but also a treatment of large deviations. This is based on Freidlin's theory of large deviations (Freidlin 1978, Freidlin and Wentzell 1984) developed in connection with the averaging principle. The latter especially concerns the approximation of the solution x^ε of a (deterministic or) stochastic differential equation

$$\dot{x}^\varepsilon(t) = b(x^\varepsilon(t), \xi(t/\varepsilon)), \quad 0 \leq t \leq T < \infty, \varepsilon > 0,$$

with $x^\varepsilon(0) = x$ in $\mathbb{R}^k$ by the solution $\bar{x}$ of the differential equation

$$\dot{\bar{x}}(t) = \bar{b}(\bar{x}(t)), \quad 0 \leq t \leq T,$$

with $\bar{x}(0) = x$, where $b : \mathbb{R}^k \times \mathbb{R}^l \to \mathbb{R}^k$ is bounded and Lipschitz continuous and ξ a (deterministic or) random bounded function on $\mathbb{R}_+$ with state space $\mathbb{R}^l$ satisfying

$$\underset{y \in \mathbb{R}^k}{\forall} \quad \frac{1}{T'} \int_0^{T'} b(y, \xi(s)) ds \to \bar{b}(y) \quad (T' \to \infty)$$

[a.s. or in probability]. For this situation the following result is due to Freidlin (1978). Assume existence of a continuous function $H : \mathbb{R}^k \times \mathbb{R}^k \to \mathbb{R}$ such that for each $x \in \mathbb{R}^k$ the function $H(x, \cdot)$ is continuously differentiable and for all piecewise constant functions $\varphi, \alpha : [0, T] \to \mathbb{R}^k$ the relation

$$\lim_{\varepsilon \to +0} \varepsilon \log E \exp \left(\int_0^{T/\varepsilon} \langle \alpha(\varepsilon u), b(\varphi(\varepsilon u), \xi(u)) \rangle du \right)$$

$$= \int_0^T H(\varphi(u),\,\alpha(u))du$$

holds. Let

$$L(v,\beta) := \sup_{z \in \mathbb{R}^k}\, (\langle \beta, z \rangle - H(v,z)),\, v \in \mathbb{R}^k, \beta \in \mathbb{R}^k,$$

and, for $\phi : [0,T] \to \mathbb{R}^k$,

$$S(T,\phi) := \int_0^T L(\phi(u),\dot{\phi}(u))du,$$

if ϕ is absolutely continuous, $S(T,\phi) := \infty$ else. Then for each $A \subset C_x[0,T]$ (linear space of continuous functions $\phi : [0,T] \to \mathbb{R}^k$ with $\phi(0) = x$, endowed with max-norm), it holds

$$- \inf_{\phi \in A^0} S(T,\phi)$$
$$\leq \varliminf_{\varepsilon} \varepsilon \log P[x^\varepsilon \in A | x^\varepsilon(0) = x]$$
$$\leq \varlimsup_{\varepsilon} \varepsilon \log P[x^\varepsilon \in A | x^\varepsilon(0) = x]$$
$$\leq - \inf_{\phi \in \overline{A}} S(T,\phi),\ x \in \mathbb{R}^k.$$

By a modification of Freidlin's theory, Kushner (1984) treats stochastic approximation processes with various sequences of weight functions. Especially for a Robbins-Monro process (X_n) in $\mathbb{R}^k$ defined by

$$X_{n+1} := X_n - \tfrac{1}{n+1}f(X_n) + \tfrac{1}{n+1}V_n$$

with Lipschitz-continuous $f : \mathbb{R}^k \to \mathbb{R}^k$ and independent identically Gaussian distributed k-dimensional random vectors V_n with $EV_n = 0$ and positive-definite covariance matrix R, he obtains

$$- \inf_{\phi \in A^0} S^*(T,\phi)$$
$$\leq \varliminf_{n} \lambda_n \log P[X^n \in A | X^n(0) = x]$$
$$\leq \varlimsup_{n} \lambda_n \log P[X^n \in A | X^n(0) = x]$$
$$\leq - \inf_{\phi \in \overline{A}} S^*(T,\phi)$$

for $A \subset C_x[0,T]$, $x \in \mathbb{R}^k$, $0 < T < \infty$. Here X^n is defined as a stochastic process with paths in $C_{\mathbb{R}^k}(\mathbb{R}_+)$ by piecewise linear interpolation, where

$$X^n(0) = X_n,\ X^n(\tfrac{1}{n+1} + \ldots + \tfrac{1}{m}) = X_m \text{ for } m > n,$$

further, for $v \in \mathbb{R}^k$, $\beta \in \mathbb{R}^k$, $s \in [0,T]$,

$$L(v,\beta,s) = \tfrac{1}{2}e^s \langle \beta - f(v),\, R^{-1}(\beta - f(v)) \rangle (1 - e^{-T}),$$

and, for $\phi : [0,T] \to \mathbb{R}^k$,

$$S^*(T,\phi) = \int_0^T L(\phi(s),\dot{\phi}(s),s)ds,$$

if ϕ is absolutely continuous, $S^*(T,\phi) = \infty$ else,

$$\lambda_n = \tfrac{1}{n}(1 - e^{-T}).$$

As a corollary, for a bounded open set G, with piecewise continuously differentiable boundary, in the attraction domain of the asymptotically stable point ϑ of the differential equation $\dot{y} = -f(y)$, the limit relation

$$\lambda_n \log P[\tau_G^n \leq T | X^n(0) = x] \to - \inf_{\phi \in A} S^*(T, \phi) \quad (n \to \infty),$$

$0 < T < \infty$, $x \in G$, concerning the probability that parts of the tail of the process escape from G, is obtained, where

$$\tau_G^n := \inf\{t \in \mathbb{R}_+; \ X^n(t) \notin G\},$$
$$A := \{\phi \in C_x[0, T]; \ \phi(t) \notin G \text{ for some } t \leq T\}.$$

The case of more general noise and the case of optimization under constraints has been treated by Dupuis and Kushner 1985 and Dupuis (and Kushner) 1987, resp.

References

Monographs
Albert, A.E., Gardner, L.A., Jr.: Stochastic Approximation and Nonlinear Regression. M.I.T. Press; Cambridge, Mass., 1967.

Benveniste, A., Métivier, M., Priouret, P.: Stochastic Approximations and Adaptive Algorithms. Springer; Berlin, Heidelberg, New York, 1990.

Chen, Han-Fu: Recursive Estimation and Control for Stochastic Systems. Wiley; New York, London, 1985.

Duflo, M.: Méthodes récursives aléatoires. Masson; Paris, 1990.

Kushner, H.J.: Approximation and Weak Convergence Methods for Random Processes, with Applications to Stochastic Systems Theory. M.I.T. Press; Cambridge, Mass., 1984.

Kushner, H.J., Clark, D.S.: Stochastic Approximation Methods for Constrained and Unconstrained Systems. Springer; Berlin, Heidelberg, New York 1978.

Ljung, L., Söderström, T.: Theory and Practice of Recursive Indentification. M.I.T. Press; Cambridge, Mass., 1983.

Marti, K.: Approximationen stochastischer Optimierungsprobleme. Hain; Königstein/Ts., 1979.

Nevel'son, M.B., Has'minskii, R.Z.: Stochastic Approximation and Recursive Estimation. Translations of Math. Monographs, Vol. 47. American Mathematical Society; Providence. R.I., 1973/76.

Schmetterer, L.: L'Approximation Stochastique. Université de Clermont-Ferrand, 2ème ed., 1972.

Tsypkin, Ya.Z.: Adaption and Learning in Automatic Systems. Academic Press; New York, London, 1971.

Tsypkin, Ya.Z.: Foundations of the Theory of Learning Systems. Academic Press; New York, London, 1973.

Survey Articles

Fabian, V.: Stochastic approximation. Optimizing Methods in Statistics (ed. J.S. Rustagi), 439 – 470. Academic Press; New York, London, 1971.

Lai, T.L.: Stochastic approximation and sequential search for optimum. Proceedings of the Berkeley Conference in Honor of Jerzy Neyman and Jack Kiefer, Vol. II (eds. L. LeCam, R.A. Olshen), 557 – 577. Wadsworth; Monterey, Ca., 1985.

Loginov, N.V.: Methods of stochastic approximation. Automat. Remote Control 27 (1966), 706 – 728.

Ruppert, D.: Stochastic approximation. Handbook of Sequential Analysis (eds. B.K. Ghosh, P.K. Sen), 503–529. Marcel Dekker; New York, 1991.

Schmetterer, L.: Stochastic approximation. Proc. Fourth Berkeley Symp. Math. Statist. Prob., I (ed. J. Neyman), 587 – 609. Univ. of California Press; Berkeley, Los Angeles, 1961.

Schmetterer, L.: Multidimensional stochastic approximation. Multivariate Analysis, II (ed. P.R. Krishnaiah), 443 – 460. Academic Press; New York, London, 1969.

Schmetterer, L.: From stochastic approximation to the stochastic theory of optimization. 11. Steiermärk. Math. Symp., Stift Rein, 1979. Bericht Nr. 127 der Mathematisch-Statistischen Sektion im Forschungszentrum Graz.

Further references (not only on stochastic approximation)

Abdelhamid, S.N.: Transformation of observations in stochastic approximation. Ann. Statist. 1 (1973), 1158 – 1174.

Anbar, D.: On optimal estimation methods using stochastic approximation procedures. Ann. Statist. 1 (1973), 1175 – 1184.

Anbar, D.: A stochastic Newton-Raphson method. J. Statist. Planning Inference 3 (1978), 153 – 163.

Arnold, L.: Stochastische Differentialgleichungen. Oldenbourg; München, 1973.

Becker, I., Greiner, G.: On the modulus of one-parameter semigroups. Semigroup Forum 34 (1986), 185 – 201.

Berger, E.: Asymptotic behaviour of a class of stochastic approximation procedures. Probab. Th. Rel. Fields 71 (1986), 517 – 552.

Billingsley, P.: Convergence of Probability Measures. Wiley; New York, London, 1968.

Billingsley, P.: Weak Convergence of Measures: Applications in Probability. Regional Conference Series in Applied Mathematics 5. SIAM; Philadelphia, Pa., 1971.

Blum, J.R.: Multidimensional stochastic approximation methods. Ann. Math. Statist. 25 (1954), 737 – 744.

Bouton, C.: Approximation gaussienne d'algorithmes stochastiques à dynamique markovienne. Ann. Inst. Henri Poincaré -Prob. Statist. 24 (1988), 131 – 155.

Chen, G.C., Lai, T.L., Wei, C.Z.: Convergence systems and strong consistency of least squares estimates in regression models. J. Multivariate Anal. 11 (1981), 319 – 333.

Clark, D.S.: Necessary and sufficient conditions for the Robbins-Monro method. Stochastic Process. Appl. 17 (1984), 359 – 367.

Cramér, H., Leadbetter, M.R.: Stationary and Related Stochastic Processes. Wiley; New York, 1967.

Daleckii, Ju. L., Krein, M.G.: Stability of Solutions of Differential Equations in Banach Space. Translations of Mathematical Monographs, Vol. 43. American Mathematical Society; Providence, R.I., 1970/74.

Deheuvels, P.: Conditions nécessaires et suffisantes de convergence ponctuelle presque sûre et uniforme presque sûre des estimateurs de la densité. C.R. Acad. Sci. Paris Ser. A 278 (1974), 1217 – 1220.

Devroye, L.: On the pointwise and integral convergence of recursive kernel estimates of probabilty densities. Utilitas Math. 15 (1979), 113 – 128.

Dupač, V.: Stochastic approximation in the presence of trend. Czech. Math. J. 16 (91) (1966), 454 – 462.

Dupuis, P.: Large deviations analysis of reflected diffusions and constrained stochastic approximation algorithms in convex sets. Stochastics 21 (1987), 63 – 96.

Dupuis, P., Kushner, H.J.: Stochastic approximations via large deviations: asymptotic properties. SIAM J. Control Optimization 23 (1985), 675 – 696.

Dupuis, P., Kushner, H.J.: Asymptotic behavior of constrained stochastic approximations via the theory of large deviations. Probab. Th. Rel. Fields 75 (1987), 223 – 244.

Dvoretzky, A.: On stochastic approximation. Proc. Third Berkeley Symp. Math. Statist. Prob., I (ed. J. Neyman), 39 – 55. Univ. of California Press; Berkeley, Los Angeles 1956.

Eckhaus, W.: New approach to the asymptotic theory of nonlinear oscillations and wave-propagation. J. Math. Anal. Appl. 49 (1975), 575 – 611.

Fabian, V.: On asymptotic normality in stochastic approximation. Ann. Math. Statist. 39 (1968), 1327 – 1332.

Fabian, V.: Asymptotically efficient stochastic approximation; the RM case. Ann. Statist. 1 (1973), 486 – 495.

Fabian, V.: A local asymptotic minimax optimality of an adaptive Robbins Monro stochastic approximation procedure. Mathematical Learning Models – Theory and Algorithms (eds. U. Herkenrath, D. Kalin, W. Vogel), 43 – 49. Springer; Berlin, Heidelberg, New York, 1983.

Frees, E.W., Ruppert, D.: Estimation following a Robbins-Monro designed experiment. To appear in J. Amer. Statist. Assoc.

Freidlin, M.I.: The averaging principle and theorems on large deviations. Russian Math. Surveys 33 (1978), 117 – 176.

Freidlin, M.I., Wentzell, A.D.: Random Perturbations of Dynamical Systems. Springer; Berlin, Heidelberg, New York, 1984.

Fritz, J.: Stochastic approximation for finding local maximum of probability densities. Studia Sci. Math. Hungar. 8 (1973), 309 – 322.

Gänssler, P., Stute, W.: Wahrscheinlichkeitstheorie. Springer; Berlin, Heidelberg, New York, 1977.

Gladyshev, E.G.: On stochastic approximation. Theory Probability Appl. 10 (1965), 275 – 278.

Goldstein, L.: Minimizing noisy functionals in Hilbert space: an extension of the Kiefer-Wolfowitz procedure. J. Theor. Probab. 1 (1988), 189 – 204.

Györfi, L.: Stochastic approximation from ergodic sample for linear regression. Z. Wahrscheinlichkeitstheorie verw. Gebiete 54 (1980), 47 – 55.

Györfi, L.: Adaptive linear procedures under general conditions. IEEE Trans. Information Theory IT – 30 (1984), 262 – 267.

Härdle, W.K., Nixdorf, R.: Nonparametric sequential estimation of zeros and extrema of regression functions. IEEE Trans. Information Theory IT – 33 (1987), 367 – 372.

Hale, J.: Theory of Functional Differential Equations. Springer; Berlin, Heidelberg, New York, 1977.

Henze, E.: Lernprozesse mit zeitabhängigen Wahrscheinlichkeiten. Zeitschr. angew. Math. Mech. 46 (1966), 297 – 302.

Herkenrath, U.: On the speed of convergence of the Kiefer-Wolfowitz stochastic approximation procedure. Math. Operationsforsch. Statist., Ser. Statistics 12 (1981), 377 – 392.

Hiriart-Urruty, J.B.: Algorithms of penalization type and dual type for the solution of stochastic optimization problems with stochastic constraints. Recent Developments in Statistics (ed. J.R. Barra et al.), 183 – 219. North-Holland; Amsterdam, New York, Oxford, 1977.

Kersting, G.: Almost sure approximation of the Robbins-Monro process by sums of independent random variables. Ann. Probab. 5 (1977), 954 – 965.

Kiefer, J., Wolfowitz, J.: Stochastic estimation of the maximum of a regression function. Ann. Math. Statist. 23 (1952), 462 – 466.

Kottmann, Th.: Learning procedures and rational expectations in linear models with forecast feedback. Diss. Univ. Bonn 1990.

Kushner, H.J.: Stochastic approximation algorithms for the local optimization of functions with nonunique stationary points. IEEE Trans. Automatic Control AC-17 (1972), 646 – 654.

Kushner, H.J.: Asymptotic behavior of stochastic approximation and large deviations. IEEE Trans. Automatic Control AC-29 (1984), 984 – 990.

Kushner, H.J., Huang, H.: Rates of convergence for stochastic approximation type algorithms. SIAM J. Control Optim. 17 (1979), 607 – 617.

Kushner, H.J., Sanvicente, E.: Stochastic approximation for constrained systems with observation noise on the system and constraints. Automatica 11 (1975), 375 – 380.

Lai, T.L., Robbins, H.: Limit theorems for weighted sums and stochastic approximation processes. Proc. Nat. Acad. Sci. USA 75 (1978), 1068 – 1070.

Lai, T.L., Robbins, H.: Consistency and asymptotic efficiency of slope estimates in stochastic approximation schemes. Z. Wahrscheinlichkeitstheorie verw. Geb. 56 (1981), 329 – 360.

Ljung, L.: Analysis of recursive stochastic algorithms. IEEE Trans. Automatic Control AC-22 (1977), 551 – 575.

Ljung, L.: Strong convergence of a stochastic approximation algorithm. Ann. Statist. 6 (1978), 680 – 696.

Mark, G.: Log-log-Invarianzprinzipien für Prozesse der stochastischen Approximation. Mitteilungen Math. Sem. Giessen 153 (1982).

McLeish, D.L.: Functional and random central limit theorems for the Robbins-Monro process. J. Appl. Probab. 13 (1976), 148 – 154.

Métivier, M., Priouret, P.: Applications of a Kushner and Clark lemma to general classes of stochastic algorithms. IEEE Trans. Information Theory IT-30 (1984), 140 – 151.

Métivier, M., Priouret, P.: Théorèmes de convergence presque sûre pour une classe d'algorithmes stochastiques à pas décroissant. Probab. Th. Rel. Fields 74 (1987), 403 – 428.

Milnor, J.W.: Topology from the Differentiable Viewpoint. Univ. Press of Virginia; Charlottesville, 1965/72.

Mohr, M.: Asymptotic theory for ordinary least squares estimators in regression models with forecast-feedback. Diss. Univ. Bonn 1990.

Nazin, A.V., Polyak, B.T., Tsybakov, A.B.: Passive stochastic approximation. Automat. Remote Control 50 (1989), 1563 – 1569.

Nixdorf, R.: An invariance principle for a finite dimensional stochastic approximation method in a Hilbert space. J. Multivariate Analysis 15 (1984), 252 – 260.

Pakes, A.: Some remarks on the paper by Theodorescu and Wolff: " Sequential estimation of expectations in the presence of trend", Austral. J. Statist. 24 (1982), 89 – 97.

Parthasarathy, K.R.: Probability Measures on Metric Spaces. Academic Press; New York, London, 1967.

Parzen, E.: On estimation of a probabilty density function and mode. Ann. Math. Statist. 33 (1962), 1065 - 1076.

Pechtl, A.: Ein Invarianzprinzip zu einem Gaußschen Markoff-Prozeß. Diplomarbeit Univ. Stuttgart 1988.

Pflug, G.: Optimale sequentielle Zerlegung. Math. Operationsforsch. Statist., Ser. Statistics 11 (1980), 287 – 295.

Pflug, G.: On the convergence of a penalty-type stochastic approximation procedure. J. Information & Optimization Sciences 2 (1981), 249 – 258.

Polyak, B. T.: New method of stochastic approximation type. Automat. Remote Control 51 (1990), 937–946.

Polyak, B. T., Juditsky, A. B.: Acceleration of stochastic approximation by averaging. Technical Report, Institute for Control Sciences of USSR Acad. Sci. (1990).

Prakasa Rao, B. L. S.: Nonparametric Functional Estimation. Academic Press; New York, London, 1983.

Révész, P.: The Laws of Large Numbers. Academic Press; New York, London. Akadémiai Kiadó; Budapest, 1968.

Révész, P.: Robbins-Monro procedure in a Hilbert space and its application in the theory of learning processes I. Studia Sci. Math. Hungar. 8 (1973), 391 – 398.

Révész, P.: Robbins-Monro procedure in a Hilbert space II. Studia Sci. Math. Hungar. 8 (1973), 469 – 472.

Révész, P.: How to apply the method of stochastic approximation in the nonparametric estimation of a regression function. Math. Operationsforschung Statist., Ser. Statistics 8 (1977), 119 – 126.

Robbins, H., Monro, S.: A stochastic approximation method. Ann. Math. Statist. 22 (1951), 400 – 407.

Robbins, H., Siegmund, D.: A convergence theorem for nonnegative almost supermartingales and some applications. Optimizing Methods in Statistics (ed. J.S. Rustagi) 233 – 257. Academic Press; New York, London, 1971.

Rockafellar, R.T.: A dual approach to solving nonlinear programming problems by unconstrained optimization. Math. Progr. 5 (1973), 354 – 373.

Rosenblatt, M.: Remarks on some nonparametric estimates of a density function. Ann. Math. Statist. 27 (1956), 832 – 837.

Ruppert, D.: Almost sure approximations to the Robbins-Monro and Kiefer-Wolfowitz processes with dependent noise. Ann. Probab. 10 (1982), 178 – 187.

Ruppert, D.: Efficient estimators from a slowly convergent Robbins-Monro process. Technical Report No. 781 (1988), School of Operations Research and Industrial Engineering, Cornell University Ithaca, New York.

Sanchez-Palencia, E.: Méthode de centrage - estimation de l'erreur et comportement des trajectoires dans l'espace des phases, Int. J. Non-Linear Mechanics 11 (176) (1976), 251–263.

Sanders, J.A., Verhulst, F.: Averaging Methods in Nonlinear Dynamical Systems. Springer; Berlin, Heidelberg, New York, 1985.

Schmetterer, L.: Sur quelques résultats asymptotiques pour le processus de Robbins-Monro. Annales Scientifiques de l'Université de Clermont 58 (1976), 166 – 176.

Schwabe, R.: Strong representation of an adaptive stochastic approximation procedure. Stoch. Processes Appl. 23 (1986), 115 – 130.

Shwartz, A., Berman, N.: Abstract stochastic approximations and applications. Stoch. Processes Appl. 31 (1989), 133 – 149.

Venter, J.H.: An extension of the Robbins-Monro procedure. Ann. Math. Statist. 38 (1967), 181 – 190.

Walk, H.: An invariance principle for the Robbins-Monro process in a Hilbert space. Z. Wahrscheinlichkeitstheorie verw. Gebiete 39 (1977), 135 – 150.

Walk, H.: Stochastic iteration for a constrained optimization problem. Commun. Statist. - Sequential Analysis 2 (1983 - 84), 369 – 385.

Walk, H.: Limit behaviour of stochastic approximation processes. Statistics & Decisions 6 (1988), 109 – 128.

Walk, H., Zsidó, L.: Convergence of the Robbins-Monro method for linear problems in a Banach space. J. Math. Anal. Appl. 139 (1989), 152 – 177.

Wei, C.Z.: Multivariate adaptive stochastic approximation. Ann. Statist. 15 (1987), 1115 – 1130.

Wertz, W.: Sequential and recursive estimators of the probability density. Statistics 16 (1985), 277 – 295.

Widrow, B., Hoff, , M.E., Jr.: Adaptive switching circuits. IRE WESCON Convention Record, part 4 (1960), 96 – 104.

Wolverton, C.T., Wagner, T.J.: Recursive estimates of probability densities. IEEE Trans. Systems Sci. Cybernet. SSC-5 (1969), 246 – 247.

Woodroofe, M.: Normal approximation and large deviations for the Robbins-Monro process. Z. Wahrscheinlichkeitstheorie verw. Geb. 21 (1972), 329–338.

Yamato, H.: Sequential estimation of a continuous probability density function and mode. Bull. Math. Statist. 14 (1971), 1 – 12.

Yosida, K.: Functional Analysis. 2nd ed. Springer; Berlin, Heidelberg, New York, 1968.

II Applicational aspects of stochastic approximation

Georg Pflug
University of Vienna
Institute of Statistics and Computer Science
Universitätsstrasse 5
A-1010 Wien
Austria

§ 7 Markovian stochastic optimization and stochastic approximation procedures

Let $F(x)$ be a real function defined on $\mathbb{R}^k$ or a subset of it. In this part we will consider the optimization problem

$$(P) \qquad \left\| \begin{array}{l} F(x) = \min! \\ x \in S \end{array} \right.$$

where $S \subseteq \mathbb{R}^k$ is a set of constraints. Any point x^* which is the solution of (P) is called a *global minimizer* of F on S. If there is an open set U such that a point x^0 is the solution of

$$(P^0) \qquad \left\| \begin{array}{l} F(x) = \min! \\ x \in S \cap U \end{array} \right.$$

then x^0 is called a *local minimizer* of F on S. In general, for deterministic procedures which use the gradient $f(x)$ of $F(x)$, only convergence to the set of critical points $\{x : f(x) = 0\}$ can be proved. There are however tricky deterministic methods which avoid convergence to non-global minimizers (Dixon and Szegö 1975; Ge 1990).

We treat only stochastic optimization procedures here. The word "stochastic optimization procedure" stands for two different situations:

(i) **Randomness in the method:** Such procedures use artificially introduced random noise, mostly to allow the search point to escape from local minima.

(ii) **Randomness in the model:** In these cases the objective function $F(x)$ as well as its derivative $f(x)$ are only observable together with some random noise. Some applications, like recursive statistical estimation, statistical pattern recognition, control and optimization of stochastic systems will be discussed in §10.

We begin with considering the simplest procedure of type (i), the random search method. This method does not make use of gradients, is universally applicable, but low in speed of convergence.

Global random search uses the fact, that an i.i.d. random sequence will hit eventually every set of positive measure. Let (η_i) be a sequence of i.i.d. random variables with a distribution ν which has support S. The algorithm keeps the best point found so far until a better point is detected:

> 1. $X_1 := \eta_1$; $n := 1$
> 2. If $F(\eta_{n+1}) < F(X_n)$
>   then $X_{n+1} := \eta_{n+1}$
>   else $X_{n+1} := X_n$.
>   Set $n := n + 1$ and go to 2.

The convergence properties of this algorithm are contained in the following theorem.

7.1. Theorem. Let $\gamma = \mathrm{ess\ inf}\, F(\eta_i) := \inf\{\beta : \nu\{F(\eta_i) \leq \beta\} > 0\}$ and suppose that $\gamma > -\infty$. Then, with the above algorithm,

$$F(X_n) \longrightarrow \gamma \text{ almost surely.}$$

PROOF: $\nu\{F(\eta_i) < \gamma\} = 0$ and therefore $\liminf_n F(X_n) \geq \gamma$ with probability one. Since $F(X_n)$ is monotonically decreasing, $\lim_n F(X_n)$ exists a.s. Let $A(\epsilon) := \{x : F(x) \leq \gamma + \epsilon\}$. By the definition of $\gamma, \nu(A(\epsilon)) > 0$ for all $\epsilon > 0$. Therefore $\nu\{\eta_i \notin A(\epsilon) \text{ for all } i\} = \prod_{i=1}^{\infty}(1 - \nu(A(\epsilon))) = 0$ and consequently $\lim_n F(X_n) \leq \gamma + \epsilon$ a.s. for all $\epsilon > 0$. Since ϵ is arbitrary, the result follows. $\square$

Although this algorithm is widely applicable (due to the weak assumptions) it has several drawbacks: Typically the sequence of the arguments (X_n) does not converge, the algorithm becomes more and more inefficient when n is large and therefore the speed of convergence is low. The exact convergence rate can be seen from the next theorem:

7.2. Theorem. Let $G(u) = \nu\{F(\eta_i) \leq u\}$. Suppose that $\lim_{x\downarrow 0} \frac{G(\gamma+kx)}{G(\gamma+x)} = k^\alpha$ for every $k > 0$. Then $a_n^{-1}(F(X_n) - \gamma)$ has a limiting distribution with distribution function $1 - \exp(-x^\alpha)$ for $x \geq 0$. a_n may be chosen as $G^{-1}(\frac{1}{n}) - \gamma$. (In most cases $a_n \sim n^{-\frac{1}{\alpha}}$.)

PROOF: This is one part of Gnedenko's famous result about limiting distributions of extremes, since $F(X_n) = \min_{i\leq n} F(\eta_i)$ (see David (1970),p. 260).
$\square$

A random search technique is called *local*, if the distribution of the search point η_n depends on the past and is typically concentrated around X_n, the best point found so far. Local search techniques need extra regularity assumptions to provide convergence, but if they do they usually converge faster. Well known procedures are due to Matyas (1965), Solis/Wets (1981) or Marti (1980).

If the objective function $F(\cdot)$ is differentiable and its gradient $f(\cdot)$ is available, it is better to use this information than to do pure random search. For to find the unconstrained minimum of $F(\cdot)$ we may go step by step in the direction of the negative gradient

$$x_{n+1} = x_n - a_n f(x_n). \tag{1}$$

This recursion was studied for the first time by v. Mises and Pollaczek-Geiringer (1929). Under mild conditions, we may establish that $F(x_n)$ is convergent and $f(x_n) \to 0$ just by taking the special case of vanishing error terms in Theorem 1.2. Notice also that it is not necessary that $a_n \to 0$ for this deterministic case. A small, but fixed a is sufficient.

The stochastic analogon of (1) is the Robbins-Monro procedure

$$X_{n+1} = X_n - a_n f(X_n) - a_n W_n \tag{2}$$

where W_n are random error variables ($\mathbb{E}(W_n) = 0$). In the first part, pointwise and functional limit theorems were discussed under various conditions about the error process W_n. In this part, we assume that W_n is conditionally independent of $X_0, \ldots, X_{n-1}$ given X_n, an assumption which guarantees that (X_n) is a Markov process.

This process is inhomogeneous, since the stepsizes a_n may vary from step to step. However, a good insight may be found by considering first the homogenous case ($a_n \equiv a$) and then go over to the inhomogenous one.

Let

$$X^a_{n+1} = X^a_n - a f(X^a_n) - a W_n \tag{3}$$

be the fixed–stepsize process and denote by $P^a = P^a(x, A)$ its *transition operator*. P^a acts from the right on probability measures μ and from left on bounded measurable functions ψ:

$$(\mu P^a)(A) := \int P^a(x, A)\, d\mu(x) \tag{4}$$

$$(P^a \psi)(x) := \int \psi(y)\, P^a(x, dy).$$

Two transitions may be composed

$$(P^a \cdot P^b)(x, A) := \int P^b(y, A)\, P^a(x, dy).$$

If X_n has distribution μ, then X_{n+1} has distribution μP^a and X_{n+2} has distribution $\mu(P^a)^2$. Let d be some distance on $\mathbb{R}^k$ and let $\mathcal{P}_d$ be the set of all

Borel measures μ on $\mathbb{R}^k$ with $\int d(x,0)\,d\mu(x) < \infty$. We introduce on $\mathcal{P}_d$ the Wasserstein-metric

$$d(\mu,\nu) = \sup\left\{\int \psi d\mu - \int \psi d\nu : \psi \text{ is Lipschitz}(1): |\psi(x) - \psi(y)| \le d(x,y)\right\}.$$

There is no ambiguity to denote the original metric d on $\mathbb{R}^k$ and the Wasserstein-metric by the same symbol, since for point masses δ_x and δ_y we have

$$d(\delta_x, \delta_y) = d(x,y). \tag{5}$$

By duality, this metric may be equivalently defined as

$$d(\mu,\nu) \quad = \quad \inf\{\mathbb{E}\,(d(X,Y)) : (X,Y) \text{ has a bivariate distribution}$$
$$\text{with marginals } \mu \text{ resp. } \nu.\}$$

(This equivalence is due to Kantorovic, see Rachev(1984)). The metric d is the biggest one which extends d by (5).

It is known that d metrizises the weak topology restricted to $\mathcal{P}_d$.

Let us introduce the coefficient of ergodicity $\rho(P)$ associated with a Markov transition P and a metric d:

$$\rho(P) := sup_{\mu \ne \nu} \frac{d\,(\mu P, \nu P)}{d(\mu,\nu)}. \tag{6}$$

By the just mentioned extremal property, we may equivalently write

$$\rho(P) := sup_{x \ne y} \frac{d\,(\delta_x P, \delta_y P)}{d(x,y)}.$$

We call P an *ergodic transition*, if there is a metric d such that

$$\rho(P) < 1.$$

For finite Markov Chains, this definition coincides with the usual notion of ergodicity (see Schachermayer and Pflug (1992)). By Banach's fixed point theorem, an ergodic P must have a unique stationary law μ^*

$$\mu^* P = \mu^*.$$

In the following, we assume that $(P^a)_{a \ge 0}$ is a family of ergodic Markov transitions with pertaining stationary laws (μ^a). For short, we write $\rho(a)$ instead of $\rho(P^a)$ whenever no confusion may occur.

In all applications, $\rho(a) \to 1$ as $a \to 0$ and the "limiting transition" P^0 is not ergodic. Let M^0 be the set of fixed points of P^0

$$M^0 = \{\mu : \mu P^0 = \mu\}. \tag{7}$$

Although M^0 may be very big, there may exist a unique weak limit

$$\mu^0 = \lim_{a \to 0} \mu^a \tag{8}$$

with $\mu^0 \in M^0$. It is the goal of recursive optimization algorithms to approach μ^0 by iteration. This can be achieved by decreasing the stepsize parameter a at each step. The decrease of stepsize must be slow enough; decreasing too fast may result in convergence to another limit than μ^0. A precise statement is given in the following Lemma.

7.3. Lemma. (Föllmer 1988). Suppose that (a_n) is a sequence of stepsizes satisfying

   (i) $a_n \to 0$,
   (ii) $\prod_n \rho(a_n) = 0$,
   (iii) $\sum_n d(\mu^{a_n}, \mu^{a_{n+1}}) < \infty$.

Then, for each starting distribution ν,

$$d(\nu \prod_{k=1}^{n} P^{a_k}, \mu^0) \longrightarrow 0 \text{ as } n \longrightarrow \infty.$$

PROOF: Let $\gamma_n = d(\mu^{a_n}, \mu^{a_{n+1}})$ and $\delta_n = d(\nu \prod_{k=1}^{n} P^{a_k}, \mu^{a_n})$.

We have

$$\begin{aligned}
\delta_{n+1} &= d\left(\nu \prod_{k=1}^{n+1} P^{a_k}, \mu^{a_{n+1}} P^{a_{n+1}}\right) \leq \\
&\leq \rho(a_{n+1})\left[d\left(\nu \prod_{k=1}^{n} P^{a_k}, \mu^{a_n}\right) + d\left(\mu^{a_{n+1}}, \mu^{a_n}\right)\right] = \\
&= \rho(a_{n+1})[\delta_n + \gamma_n] \, .
\end{aligned}$$

Consequently

$$\delta_n \leq \left[\prod_{k=2}^{n} \rho(a_k)\right]\delta_1 + \sum_{k=1}^{n-1} \gamma_k \prod_{\ell=k+1}^{n} \rho(a_\ell) \longrightarrow 0$$

and also

$$d\left(\nu \prod_{k=1}^{n} P^{a_k}, \mu^0\right) \leq \delta_n + d(\mu^{a_n}, \mu^0) \longrightarrow 0.$$

$\square$

Markov transitions are not commutative in general and we agree that $\prod P^{a_k}$ means that the product is in ascending order of k.

7.4. Remark. Lemma 7.3 remains valid, if we replace (ii) by the weaker condition (ii') there is a sequence of ascending integers $0 = l_1 < l_2 < \ldots$ such that

$$\prod_j \rho\left(\prod_{i=l_j+1}^{l_{j+1}} P^{a_i}\right) = 0.$$

The proof follows the same lines as the proof of Lemma 7.3 and is therefore omitted.

Condition (ii') is indeed weaker than (ii) since the coefficient of ergodicity is submultiplicative

$$\rho(P \cdot Q) \leq \rho(P) \cdot \rho(Q)$$

and hence

$$\bullet \quad \prod_{j=1}^{\infty} \rho\left(\prod_{i=l_j+1}^{l_{j+1}} P^{a_i}\right) \leq \prod_{i=1}^{\infty} \rho(P^{a_i}).$$

Lemma 7.3 has important applications in discrete global optimization and Simulated Annealing (Dekker and Aarts (1988)). We pursue here the application for the Robbins Monro procedure. The first result proves that a unique stationary μ^a exists for the process (3) under some regularity assumptions.

7.5. Assumption.

(i) there is a compact set C such that
$$||x - af(x)|| < (1 - a\lambda_0)||x||$$
for some $\lambda_0 > 0$, sufficiently small a and all $x \notin C$,

(ii) $x \mapsto ||f(x)||$ is upper semicontinuous,

(iii) (W_n) is an i.i.d. sequence with density $g(y)$ satisfying
$\inf\{g(x) : ||x|| \leq K\} > 0$ for all $K > 0$ and $\int ||x||^3 g(x)dx < \infty$.

7.6. Lemma. Let the Assumption 7.5 be satisfied. Then the processes (X_n^a) are recurrent in the sense of Harris (Revuz (1975), p.75) and their unique invariant measures μ^a satisfy

$$\sup_a \int ||x||^3 d\mu^a(x) < \infty.$$

In particular, the measures (μ^a) are uniformly tight.

PROOF: For simplicity, we write X_n instead of X_n^a in this proof. Without loss of generality we may assume that $C = \{x : ||x||^2 \leq \gamma\}$ for a γ s.t.

$$\gamma > \frac{a^2 E(||W_n||^2)}{a\lambda_0(1 - a\lambda_0)}. \tag{9}$$

Consider the following stopping times

$$\begin{aligned}
\sigma_1 &= \inf\{n : X_n \notin C\} \\
\tau_i &= \inf\{n > \sigma_i : X_n \in C\} \\
\sigma_i &= \inf\{n > \tau_{i-1} : X_n \notin C\}.
\end{aligned}$$

We show that $\tau_i - \sigma_i$ is a. s. finite for all i. By the strong Markov property we may assume that $i = 1$ and $\sigma_i = 1$. Let $\mathcal{F}_n$ be the σ-algebra generated by $(X_1, \ldots, X_n)$. On the set $\{\tau_1 > n\} \in \mathcal{F}_n$ we have by (9)

$$
\begin{aligned}
\mathbb{E}(\|X_{n+1}\|^2 | \mathcal{F}_n) &\leq \|X_n - af(X_n)\|^2 + a^2 \mathbb{E}(\|W_n\|^2) \\
&\leq (1 - a\lambda_0)^2 \|X_n\|^2 + a^2 \mathbb{E}(\|W_n\|^2) \\
&\leq (1 - a\lambda_0)\|X_n\|^2.
\end{aligned}
$$

It follows that $V_n := \|X_{n \wedge \tau_1}\|^2 \cdot (1 - a\lambda_0)^{-(n \wedge \tau_1)}$ is a nonnegative supermartingale. Since a supermartingale satisfies

$$
\mathbb{P}_{V_1}\{V_n \geq b\} \leq \min\left(\frac{V_1}{b}, 1\right)
$$

(Neveu (1974), prop. 11-2-7, $\mathbb{P}_{V_1}$ denotes the conditional probability given V_1) we conclude that

$$
\mathbb{P}_{X_1}\{\tau_1 > k\} \leq \mathbb{P}_{X_1}\{\|X_{k \wedge \tau_1}\|^2 > \gamma\}
$$
$$
\leq \mathbb{P}_{X_1}\left\{V_k > \gamma \cdot (1 - a\lambda_0)^{-k}\right\} \leq \min\left\{\gamma^{-1}(1 - a\lambda_0)^{k-1}\|X_1\|^2, 1\right\}
$$

and, in particular,

$$
\mathbb{P}_{X_1}\{\tau_1 = \infty\} = \lim_{k \to \infty} \mathbb{P}_{X_1}\{\tau_1 > k\} = 0
$$

for all $a > 0$. Consequently $\mathbb{P}_{X_1}\{X_n \in C \text{ infinitely often}\} = 1$. Let A be any open set. We have to show that the process visits A infinitely often for every starting value x, i.e.

$$
\mathbb{E}\left(\sum_{n=1}^{\infty} 1_A(X_n) | X_1 = x\right) = \infty.
$$

By assumption (iii)

$$
\inf_{y \in C} \mathbb{E}(1_A(X_{i+1}) | X_i = y) > 0
$$

and hence

$$
\mathbb{E}\left(\sum_{\{X_n \in C\}} 1_A(X_{n+1}) | X_1 = x\right) = \infty
$$

since $\{X_n \in C\}$ infinitely often.
Using the inequality

$$
\|X_{n+1}\|^3 \leq \|X_n - af(X_n)\|^3 + 6 \cdot a \cdot \|X_n - af(X_n)\|^2 \cdot \|W_n\| + 6a^3 \|W_n\|^3
$$

and (i) it may easily be shown that $\mathbb{E}\left(\|X_{(\sigma_i + n) \wedge \tau_i}\|^3\right)$ is decreasing for sufficiently small a and suitable γ and since
$\mathbb{E}\left(\|X_{(\tau_i + n) \wedge \sigma_{i+1}}\|^3\right)$ is bounded by (ii) and (iii) it follows that

$$
\limsup_{n \to \infty} \mathbb{E}\left(\|X_n\|^3\right) < \infty
$$

uniformly in a and therefore

$$\sup_a \int ||x||^3 d\mu^a(x) < \infty.$$

$\square$

The explicit calculation of μ^a requires the solution of an integral equation, which has no closed form in general. It is however possible to derive a comparison result which allows to find bounds for the unknown μ^a. For the univariate case this was done by Högnäs (1986) and generalized to the multidimensional case by the author (1991).

We prove here only that μ^a becomes more and more concentrated around the critical points of F, as a tends to zero. Only the univariate case will be considered, but the multivariate generalization is easy.

7.7. Lemma. Let μ^a be the stationary distribution of

$$X^a_{n+1} = X^a_n - af(X^a_n) - aW_n. \tag{10}$$

Let Assumption 7.5 be satisfied and suppose that f is continuous. Then, as $a \to 0$,

$$\mu^a(A) \to 1$$

for all open sets A containing the set of critical points $B = \{x : f(x) = 0\}$.

PROOF: The Assumption 7.5 guarantees that B is a bounded set. Let $A \supseteq B$ be an open set and K a large constant such that $B \subseteq [-K, K]$. We may construct a twice differentiable function ψ with the following properties

(i) $\text{sign}(\psi'(x)) = - \text{sign}(f(x))$ inside $[-2K, 2K]$
(ii) $\psi(x) = 0$ outside $[-2K, 2K]$
(iii) $-f(x) \cdot \psi'(x) \geq 1$ in $[-K, K] \backslash A$
(iv) $\sup_x \psi''(x) < \infty$.

Let (X^a_n) be stationary. We omit the superscript a further on and get by a Taylor expansion

$$\begin{aligned}
\psi(X_{n+1}) &= \psi(X_n - af(X_n) - aW_n) \\
&= \psi(X_n) - (af(X_n) + aW_n)\psi'(X_n) \\
&+ \frac{1}{2}(af(X_n) + aW_n)^2 \psi''(\tilde{X}_n)
\end{aligned}$$

for some point $\tilde{X}_n$ lying between X_n and X_{n+1}.
Taking the expectation on both sides and using the stationarity we get

$$0 = -a\mathbb{E}(f(X_n)\psi'(X_n)) + O(a^2).$$

Consequently, dividing by a,

$$\mathbb{E}(-f(X_n)\psi'(X_n)) = O(a). \tag{11}$$

$-f(x)\psi(x)$ is a nonnegative function which is greater or equal $\mathbf{1}_{[-K,K]\setminus A}(x)$. By (11), $\mu^a([-K,K]\setminus A) \to 0$. Since K is arbitrary and the (μ^a) are uniformly tight, $\mu^a(A) \to 1$.

$\square$

The recursion (3) cannot distinguish between global and local minima of F. If one wants to guarantee that only global minimizers are limit points, one has to increase the variance of the error term considerably in order to allow an escape from local minima. We start with replacing aW_n by $\sqrt{a}W_n$ in (10) and consider

$$X_{n+1} = X_n - af(X_n) - \sqrt{a}W_n. \tag{12}$$

We assume that f satisfies the following assumption:

7.8. Assumption.
  (i) There is a $K > 0$ and a $k_1 > 0$ such that $\text{sign}\,(x)\cdot f(x) \geq k_1$ for $|x| \geq K$,
 (ii) $|f(x)| \leq k_2\cdot|f(u)|$ for $K < |x| < |u|$,
 (ii) $|f'(x)| \leq k_3$.

This assumption implies that $F(x) \to \infty$ as $|x| \to \infty$ and that

$$\int \exp\left(-\frac{2F(u)}{\sigma^2}\right)\,du < \infty$$

for all $\sigma^2 > 0$.

7.9. Lemma. Let ν^a be the stationary distribution of

$$X_{n+1} = X_n - af(X_n) - \sqrt{a}W_n$$

and let ν^0 be the probability measure with density

$$\text{const.}\exp\left(-\frac{2F(x)}{\sigma^2}\right),$$

where $\sigma^2 = \text{Var}\,(W_n)$.
Define $\mathcal{H}_L = \{\phi : |\phi(\cdot)| \leq L; |\phi'(\cdot)| \leq L, |\phi(\cdot)\cdot f(\cdot)| \leq L\}$ as a class of test functions. Then, under assumption 7.8,

$$\sup\left\{\int\phi d\nu^a - \int\phi d\nu^0 : \phi \in \mathcal{H}_L\right\} = O(\sqrt{a})$$

as $a \to 0$ (for all $L > 0$). In particular, ν^a converges weakly to ν^0.

PROOF: Let $\phi \in \mathcal{H}_L$. W.l.o.g. we may assume that $\int \phi \, d\nu^0 = 0$, otherwise we replace ϕ by $\phi - \int \phi \, d\nu^0$. Let

$$\psi'(x) := \int_{-\infty}^x \frac{2}{\sigma^2} \cdot \phi(u) \cdot \exp\left(\frac{2}{\sigma^2}(F(x) - F(u))\right) du.$$

ψ' is bounded: Since it is continuous, it suffices to show the boundedness for $|x| > K$. Let $x > K$ (the other case is similar). Then

$$|\psi'(x)| \le \frac{2}{\sigma^2} \exp(\frac{2F(x)}{\sigma^2}) \int_x^\infty \frac{L}{k_1} f(u) \exp(-\frac{2F(u)}{\sigma^2}) \, du = \frac{L}{k_1}.$$

Taking the derivative of ψ', we get

$$\psi''(x) = \frac{2}{\sigma^2}\phi(x) + \frac{2}{\sigma^2}\psi'(x) \cdot f(x). \tag{13}$$

Taking another derivative, we obtain

$$\psi'''(x) = \frac{2}{\sigma^2}\phi'(x) + \frac{2}{\sigma^2}\psi'(x)f'(x) + \frac{4}{\sigma^4}\phi(x)f(x) + \frac{4}{\sigma^4}\psi'(x)f^2(x). \tag{14}$$

We show that ψ''' is bounded. The first three summands in (14) are bounded by assumption. The boundedness of the last summand for $x \ge K$ follows from $f^2(x)|\psi'(x)| \le$

$$\le f^2(x)\frac{2}{\sigma^2}\exp(\frac{2F(x)}{\sigma^2}) \int_x^\infty |\phi(u)f(u)|f(u)f^{-2}(u)\exp(-\frac{2F(u)}{\sigma^2}) \, du$$

$$\le f^2(x)\frac{2}{\sigma^2}\exp(\frac{2F(x)}{\sigma^2}) \int_x^\infty Lk_2^2 f(u)f^{-2}(x)\exp(-\frac{2F(u)}{\sigma^2}) \, du = Lk_2^2.$$

The case $x \le -K$ is analogous. Let $\psi(x) = \int_0^x \psi'(u) \, du$ and (X_n) be stationary. By a Taylor expansion we get

$$\begin{aligned} \psi(X_{n+1}) &= \psi(X_n - af(X_n) - \sqrt{a}W_n) \\ &= \psi(X_n) - (af(X_n) + \sqrt{a}W_n)\psi'(X_n) + \\ &+ \frac{1}{2}(af(X_n) + \sqrt{a}W_n)^2\psi''(X_n) - \\ &- \frac{1}{6}(af(X_n) + \sqrt{a}W_n)^3(\psi'''(\tilde{X}_n)) \end{aligned}$$

for some point $\tilde{X}_n$ lying between X_n and X_{n+1}.
Taking the expectation on both sides and using the stationarity we get

$$0 = -a\, \mathbb{E}(f(X_n)\psi'(X_n)) + \frac{1}{2}a\sigma^2\, \mathbb{E}(\psi''(X_n)) + O(a^{3/2}).$$

Consequently, by (13),

$$\int \phi \, d\nu^a = \mathbb{E}(\phi(X_n)) = \mathbb{E}(-f(X)\psi'(X)) + \frac{1}{2}\sigma^2\, \mathbb{E}(\psi''(X)) = O(\sqrt{a}).$$

The class of functions $\mathcal{H}_L$ contains all trigonometric functions that are made zero outside a compact interval in a continuous way. This family is rich enough to determine weak convergence.   □

We remark that the limit distribution ν^0 has nonvanishing density and may therefore be viewed as Gibbs distribution (see definition below).

7.10. Definition. A distribution with density

$$\text{const. } \exp(-\frac{G(x)}{d}) \tag{15}$$

is called *Gibbs distribution* with *energy function G*.

The following limit result is known for Gibbs distributions:

7.11. Lemma. As d tends to 0, the Gibbs distribution converges weakly to the uniform distribution on all global minima of G.

PROOF: see Hwang(1980).   □

In order to make use of the limiting property indicated in Lemma 7.11 the variance σ^2 should be decreased in (12) gradually. We modify the recursion again and consider

$$X_{n+1} = X_n - af(X_n) - \sqrt{a \cdot \delta(a)} \cdot W_n$$

for $\delta(a)$ tending to zero as $a \to 0$. Introduce an additional assumption.

7.12. Assumption.

 (i) $F(x) \geq 0$ and $F(x) = 0$ for the global minimizers (this is no loss of generality)
 (ii) There are finitely many global minimizers $\xi_1, \ldots, \xi_k$ of F.

7.13. Lemma. Let γ^a be the stationary distribution of

$$X_{n+1} = X_n - af(X_n) - \sqrt{a \cdot \delta(a)} \cdot W_n. \tag{16}$$

Let Assumption 7.12 be fulfilled. If

$$\delta(a) = \frac{1}{\log \log(1/a)}, \tag{17}$$

then all weak limits of γ^a as $a \to 0$ are concentrated on the set of global minimizers of F.

PROOF: Let $\tilde{\gamma}^a$ be the distribution with density

$$\text{const. } \exp\left(-\frac{2F(x)}{\sigma^2 \delta(a)}\right).$$

Fix an $\epsilon > 0$ and a large constant K such that $F(x) > 2\epsilon$ for $|x| > K$. Let

$$\phi_1(x) = \mathbf{1}_{\{F(x) \geq 3\epsilon\} \cap [-K,K]}$$

$$\phi_2(x) = \mathbf{1}_{\{\epsilon \leq F(x) \leq 2\epsilon\}}$$

and

$$\phi_a(x) = \phi_1(x) - \eta(a) \cdot \phi_2(x).$$

Choose $\eta(a)$ such that

$$\int \phi_a(x) \, d\tilde{\gamma}^a(x) = 0.$$

Let $\lambda(\epsilon)$ be the Lebesgue measure of the set $\{\epsilon \leq F(x) \leq 2\epsilon\}$. Since

$$\eta(a) = \frac{\int \phi_1 \, d\tilde{\gamma}^a}{\int \phi_2 \, d\tilde{\gamma}^a} \leq \frac{2K \exp(-6\epsilon\sigma^{-2}\delta(a)^{-1})}{\lambda(\epsilon) \exp(-2\epsilon\sigma^{-2}\delta(a)^{-1})}$$

and $\delta(a) \to 0$, it follows that

$$\eta(a) \to 0 \text{ as } a \to 0.$$

Consider the function

$$\psi_a'(x) = \int_{-\infty}^x \frac{2}{\sigma^2 \cdot \delta(a)} \cdot \phi_a(u) \cdot \exp\left(\frac{2}{\sigma^2 \cdot \delta(a)}(F(x) - F(u))\right) du.$$

If $x \geq K$ then by assumption 7.8 (i)

$$\psi_a'(x) = \int_x^\infty \frac{2}{\sigma^2 \cdot \delta(a)} \cdot \phi_a(u) \cdot \exp\left(\frac{2}{\sigma^2 \cdot \delta(a)}(F(x) - F(u))\right) du$$

$$\leq \quad \text{const.} \int_x^\infty \frac{2f(u)}{\sigma^2 \cdot \delta(a)} \exp\left(\frac{2}{\sigma^2 \cdot \delta(a)}(F(x) - F(u))\right) du \leq \quad \text{const.}$$

Consequently ψ_a' is bounded for $x \geq K$ and the same is true for $x \leq -K$. Since F is bounded for $|x| < K$ we have

$$\sup_x |\psi_a'(x)| = O(\delta^{-1} \exp(c_1/\delta(a)))$$

for some c_1 as $a \to 0$. Notice that

$$\phi_a(x) = -f(x)\psi_a'(x) + \frac{\delta(a) \cdot \sigma^2}{2} \psi_a''(x) \tag{18}$$

and therefore

$$\sup_x |\psi_a''(x)| = O(\delta(a)^{-2} \exp(c_1/\delta(a)))$$

and

$$\sup_x |\psi_a'''(x)| = O(\delta(a)^{-3} \exp(c_1/\delta(a))). \tag{19}$$

Let $\psi_a(x) = \int_0^x \psi_a'(u)\,du$ and (X_n) be stationary. By a Taylor expansion we get

$$
\begin{aligned}
\psi_a(X_{n+1}) &= \psi_a(X_n - af(X_n) - \sqrt{a \cdot \delta(a)}W_n) \\
&= \psi_a(X_n) - (af(X_n) + \sqrt{a \cdot \delta(a)}W_n)\psi_a'(X_n) + \\
&+ \frac{1}{2}(af(X_n) + \sqrt{a \cdot \delta(a)}W_n)^2\psi_a''(X_n) + \\
&+ \frac{1}{6}(af(X_n) - \sqrt{a \cdot \delta(a)}W_n)^3(\psi_a'''(\tilde{X}_n))
\end{aligned}
$$

for some point $\tilde{X}_n$ lying between X_n and X_{n+1}.

Taking the expectation on both sides and using the stationarity and (19) we get

$$
\begin{aligned}
0 &= -a\,\mathbb{E}(f(X_n)\psi_a'(X_n)) + \frac{1}{2}a\delta(a)\sigma^2\,\mathbb{E}(\psi_a''(X_n)) \\
&+ O\left(a^{3/2}\delta(a)^{-3/2}\exp(c_1/\delta(a))\right).
\end{aligned}
$$

Consequently, dividing by a and using (17) and (18)

$$
\int \phi_a\,d\gamma^a = \mathbb{E}(\phi_a(X_n)) = \mathbb{E}(-f(X)\psi_a'(X)) + \frac{\delta(a)\sigma^2}{2}\,\mathbb{E}(\psi_a''(X)) = o(1).
$$

Since ϕ_a converges uniformly to ϕ_1, one sees that $\gamma^a(\{x\,:\,F(x) \geq 3\epsilon\} \cap [-K, K]) \to 0$. Since ϵ and K are arbitrary, all weak limit points of γ^a are concentrated on the global minimizers of F.

$\square$

Remark. As was seen in the previous proof, $\delta(a)$ should fulfill

$$
[\delta(a)]^{3/2}\exp(\frac{c_1}{\delta(a)}) = o(a^{-1/2}). \tag{20}
$$

Alternatively to the choice (17) one could also take

$$
\delta(a) = \frac{c}{\log(1/a)}, \qquad c \text{ large.}
$$

The just proved result may be put in the framework of the discussion before Lemma 7.3. Consider the recursions (10) and (16). If we set a equal to 0, then we arrive at the trivial process

$$
X_{n+1} = X_n,
$$

which has all probability measures as stationary distributions. Thus the set M^0 in (7) is the set of all probabily measures. Nevertheless, the sequence of stationary distributions of (10) converges weakly to a specific limit which sits at the critical points of f. In the case of the recursion (16) the limit sits at the global minimizers of F. Thus (8) is true in both cases (10) and (16) and the question arises whether we may apply Lemma 7.3 and design a variable step procedure with weak limit sitting on the critical points resp. on the global minimizers.

According to the cited Lemma one has to find a metric d, which leads to a contracting coefficient of ergodicity. If f has a unique root, then such a metric exists under some regularity assumptions. If however there are several roots, then the choice of the stepsize constants decides whether the set of all possible limit points is the set of all local minima of F or the set of all global minima of F. This will be discussed below.

We begin with the case of a unique root and a strong regularity assumption.

7.14. Assumption. There is a constant $\lambda_0 > 0$ such that

$$\|x - af(x) - y + af(y)\| \le (1 - a\lambda_0)\|x - y\|$$

for all x, y and sufficiently small a.

Under this assumption f has at most one root x^*. In order to apply Lemma 7.3 we have to find the appropriate metric. We may simply take

$$d(x, y) = \|x - y\|$$

here. The pertaining coefficient of ergodicity is bounded by

$$\rho(a) \le \sup_{x \neq y} \frac{\mathbb{E}(x - af(x) - aW - (y - af(y) - aW))}{\|x - y\|} = (1 - a\lambda_0).$$

Since $\prod_n (1 - a_n) = 0$ iff $\sum_n a_n = \infty$, we arrive at the following result:

7.15. Theorem. Under Assumption 7.14, if $a_n \downarrow 0$ and $\sum a_n = \infty$, then for the procedure

$$X_{n+1} = X_n - a_n \cdot f(X_n) - a_n \cdot W_n$$

it follows that

$$X_n \to x^* \text{ in probability.}$$

SKETCH OF THE PROOF: Only the condition $\sum_n d(\mu^{a_n}, \mu^{a_{n+1}}) < \infty$ remains to be checked. One shows that

$$d(\mu^a, \mu^a P^b) = O(a|b - a|)$$

and since

$$d(\mu^a, \mu^b) \le \sum_{i=0}^{\infty} d(\mu^a (P^b)^i, \mu^a (P^b)^{i+1}) \le$$

$$\le d(\mu^a, \mu^a P^b) \sum_{i=0}^{\infty} \rho^i(P^b) \le d(\mu^a, \mu^a P^b)[1 - \rho(P^b)]^{-1}$$

one sees that

$$d(\mu^{a_n}, \mu^{a_{n+1}}) = O(\frac{1}{n} \cdot |\frac{1}{n+1} - \frac{1}{n}| \cdot n) = O(n^{-2}).$$

□

Remark that the condition $\sum a_n^2 < \infty$ is not required for this conclusion, which is relatively weak compared to the stronger results of §1.

The case of non–uniqueness of the roots of $f(x) = 0$ will be treated only in the one–dimensional case. If one applies the usual Robbins–Monro (2) with $\sum a_n = \infty, \sum a_n^2 < \infty$, the following result may be proved:

7.16. Theorem. Let $B = \{x : f(x) = 0\}$ be the set of critical points and

$$\tilde{B} = \{x : f(y) \cdot \text{sign}(x - y) \geq 0 \text{ for all } y \text{ in a neighborhood of } x\}.$$

Suppose that the conditions of Theorem 1.2b are fulfilled and assume that $\text{Var}(W_n|X_n) \geq \sigma^2 > 0$. Then all limit points of the stochastic approximation procedure (2) lie in $B \backslash \tilde{B}$.

PROOF: see Nevelson and Hasminskii (1972), Theorem 4.1. The formulation there contains also the multidimensional case. The appropriate definition of $\tilde{B}$ in the mutidimensional case is

$$\tilde{B} = \{x : \quad \text{there is a positive definite symmetric matrix } A \text{ s.t.}$$
$$f(y)^t \cdot A(x - y) \geq 0 \text{ for all } y \text{ in a neighborhood of } x\}.$$

□

This result shows that one can expect that only local minima appear as limit points. In general, one cannot expect that every limit point is a global minimum. However, we may use Lemma 7.3 and Lemma 7.13 in the following way:

Introduce the discrete metric

$$d(x, y) = \begin{cases} 0 & \text{if } x \neq y \\ 1 & \text{if } x = y. \end{cases}$$

and the Wasserstein metric pertaining to it

$$d(\mu, \nu) = 1 - \int \min \left(\frac{d\mu}{d\lambda}, \frac{d\nu}{d\lambda} \right) d\lambda,$$

where λ is a dominating measure.

Suppose that all global minimizers of F are contained in a compact set $[-K, K]$, which is known. We use a "reflected" version of the Robbins–Monro procedure: Let

$$T(x) = \begin{cases} x & x \in [-K, K] \\ 2nK - x & x \in [(2n-1)K, (2n+1)K] \\ -2nK - x & x \in [-(2n+1)K, -(2n-1)K] \end{cases}$$

be the reflection function. Based on the recursion (16) we define the reflected process

$$X_{n+1} = T(X_n - af(X_n) - \sqrt{a \cdot \delta(a)} \cdot W_n). \quad \text{.} \tag{21}$$

Suppose that W_n has density $g(\cdot)$. Then the ergodic coefficient of (21) satisfies

$$\rho(a) \leq 1 - \frac{2K}{\sqrt{a\delta(a)}} \inf_{x \in [-2K, 2K]} g(x/\sqrt{a\delta(a)}).$$

The appropriate choice of a_n depends on the density g. For the normal density $g(x) = \frac{1}{\sqrt{2\pi}} \exp(-\frac{x^2}{2})$ we have

$$\rho(a) \leq 1 - \frac{c_2}{\sqrt{a\delta(a)}} \exp\left(-\frac{c_2^2}{a\delta(a)}\right)$$

for some constant c_2, for the Cauchy density $g(x) = \frac{1}{\pi} \frac{1}{1+x^2}$ however

$$\rho(a) \leq 1 - \frac{c_3 \, a \, \delta(a)}{c_3^2 + a^2 \delta^2(a)}$$

for some constant c_3. Using Lemma 7.3 we get the following result:

7.17. Theorem. Let $a_n \downarrow 0$ be chosen such that
$$a_n \delta(a_n) = \frac{1}{\log(n)} \qquad \text{(if } W_n \text{ is normally distributed)}$$
$$a_n \delta(a_n) = \frac{1}{n} \qquad \text{(if } W_n \text{ is Cauchy--distributed)}$$
where $\delta(a)$ fulfills condition (20). Then all limit points of the recursion

$$X_{n+1} = T(X_n - a_n f(X_n) - \sqrt{a_n \delta(a_n)} W_n)$$

are concentrated on the global minimizers of F in $[-K, K]$.
PROOF: A consequence of Lemma 7.3. □

One may also make use of Remark 7.4 and derive a stronger result by combining several steps of the procedure into one block. Choose $l_j = [e^j]$, where [] denotes the integer part. Then, if $|f'| \leq k_3$ and (W_n) are normally distributed, the conditional density of $X_{l_{j+1}}$ given $X_{l_j} = x$ is bounded below by

$$\frac{c_2}{\sqrt{\gamma_j}} \exp(-\frac{c_2}{\gamma_j})$$

uniformly in x, where

$$\gamma_j = \sum_{i=l_j}^{l_{j+1}} a_i \delta(a_i) \prod_{s=i}^{l_{j+1}} (1 - k_3 a_s)^2.$$

Choosing

$$a_n = \frac{1}{n}; \qquad \sqrt{a_n \delta(a_n)} = \sqrt{\frac{1}{n \log \log n}} \tag{22}$$

it follows that

$$\gamma_j \sim \frac{1}{\log(j+1)}$$

and hence by Remark 7.4 the same conclusion can be made as in Theorem 7.17.

The just presented result uses a reflection and needs the advance knowledge of a compact interval, which contains all global minimizers. A similar assumption was made by Kushner(1987). Recently Gelfand and Mitter (1991) were able to get rid of this assumption and proved a convergence result for the non–reflected procedure

$$X_{n+1} = X_n - \frac{A}{n}f(X_n) - \sqrt{\frac{B}{n \log \log n}} W_n,$$

which uses similar stepsize constants as in (22).

In the case of a unique minimizer x^*, the conclusion

$$\mu^a \to \delta_{x^*}$$

is of "Law of Large Numbers" type and relatively weak. More detailed information can either be gained by the theory of large deviations (see chapter 6) or by "blowing up" the scale around x^* to get a "Central Limit" like result: Let X^a be distributed according to μ^a. We may ask about the asymptotic distribution of

$$\gamma(a)(X^a - x^*)$$

as $a \to 0$, where $\gamma(a)$ is a normalizing constant.

7.18. Theorem. Let Assumption 7.5 be satisfied. Moreover assume that

(i) For every $\epsilon > 0$ there is a $\delta_\epsilon > 0$ such that

$$\inf_{\|x-x^*\| \geq \epsilon} \frac{(x-x^*)^t f(x)}{\|x-x^*\|^2} \geq \delta_\epsilon$$

(ii) $\|f(x)\| \leq K_1 + K_2\|x - x^*\|$ for some constants K_1, K_2

(iii) $f(x) = A \cdot (x - x^*) + o(x - x^*)$ near x^*.

Then the distribution of $\sqrt{a}(X^a - x^*)$ (X^a stationary), converges weakly to a normal $N(0, V)$ distribution, where the covariance matrix V satisfies

$$AV + VA^t = C \tag{23}$$

with C being the covariance matrix of W_n.

PROOF: see Kushner/Hai–Huang(1981) or Pflug(1986).   □

The equation (23) can be solved explicitly (see discussion after Theorem 8.5).

We close this section by comparing the discussed results for stochastic recursions with continuous–time stochastic differential equations (SDE's): Consider the SDE

$$dX(t) = -f(X(t))dt + \sigma(X(t))dW(t). \tag{24}$$

We introduce the *scale density*

$$s(x) = \exp\left(2\int_{x_0}^x \frac{f(u)}{\sigma^2(u)}\,du\right)$$

the *scale function*

$$S(x) = \int_{x_0}^x s(u)\,du$$

and the *speed density*

$$m(x) = \frac{1}{s(x)\cdot\sigma^2(x)}.$$

The following theorem characterizes the stationary distribution:

7.19. Theorem.

(a) If the constants c_1 and c_2 can be determined such that

$$g(x) = m(x)[c_1 S(x) + c_2]$$

is a probability density, then this is a stationary density of the process (24).

(b) If the process (24) is reflected at the boundaries $-K$ and K, then the stationary density of this process is

$$g(x) = \text{const. } m(x), \qquad x \in [-K, K].$$

PROOF: see Karlin/Taylor (1981), p. 221.   □

As a consequence, the stationary distribution of

$$dX(t) = -f(X(t))dt + \sqrt{2a}\,dW(t)$$

is of the Gibbs type

$$g(x) = \text{const. } \exp\left(-\frac{F(x)}{a}\right)$$

and this may be used for global optimization: If $X(t)$ fulfills

$$dX(t) = -f(X(t))dt + \sqrt{2a(t)}\,dW(t)$$

with

$$a(t) = \frac{a_1}{\log(a_2 + t)}$$

and $X(\cdot)$ is reflected at the boundaries of a finite interval, then X converges in law to the uniform distribution on the global minima of F in this interval (see Geman/Hwang (1986) and Kushner(1987)).

The multidimensional case is analoguous: By the Fokker–Plank equation, the stationary distribution of the system of equations

$$dX_i(t) = -f_i(X(t))dt + \sqrt{2a} \sum_j \sigma_{ij}(X(t))dW_j(t)$$

with W_j independent Brownian motions has a stationary distribution with density g which is the solution of

$$\sum_i \frac{\partial}{\partial x_i}(f_i(x)g(x)) + a \sum_{ij} \frac{\partial}{\partial x_i \partial x_j}(\sigma_{ij}(x)g(x))$$

(cf. Gihman and Skohorod (1968), p. 138 ff.). The explicit solution of this equation is known only if (σ_{ij}) is the unit matrix. In this case

$$g(x) = \text{const.} \exp\left(-\frac{F(x)}{a}\right)$$

which is again of Gibbs type. The multivariate case of global minimization with the use of k-dimensional Brownian motion was considered by Chiang, Hwang and Sheu (1987).

§8   Asymptotic distributions

In this section, we consider the asymptotic distribution of a recursion of the form

$$X_{n+1} = X_n - \frac{a}{n}f(X_n) + \frac{a}{n} \cdot W_n. \tag{25}$$

where $\mathbb{E}(W_n) = 0$, $\text{Var}(W_n) < \infty$. W. l. o. g. we assume that 0 is the unique root of f. It was shown by Chung(1954) that $\sqrt{n}(X_n)$ is asymptotically normal if $f'(0) = \alpha$ and $2a\alpha > 1$. He used a moment method. Other proofs are due to Sacks (1958), Fabian (1968) and Kersting (1978). Since Kersting's method works also for degenerate cases which exhibit non-normal limit laws, this method will be presented here.

We begin with studying the situation where $f'(0) = 0$, or, more precisely, where

$$\lim_{x \to 0} \frac{f(x)}{|x|^\gamma \cdot \text{sign}(x)} = \alpha$$

for a $\gamma > 1$.

Let us state the following set of assumptions.

8.1. Assumption.

(i) $x \cdot f(x) > 0 \qquad$ for $x \neq 0$

(ii) $f(x) = \alpha|x|^{\gamma} \cdot \operatorname{sgn}(x) \cdot (1 + o(1)) \qquad$ as $x \to 0$

(iii) $|f(x)| \leq A|x| + B \qquad$ for some constants $A, B \geq 0$

(iv) $\operatorname{Var}(W_n|X_n) \leq \sigma^2$

($\operatorname{Var}(W_n|X_n)$ denotes the conditional variance of W_n given X_n).

It can easily be seen that the deterministic procedure

$$x_{n+1} = x_n - \frac{a}{n} \cdot f(x_n)$$

satisfies

$$|x_n| = O\left((\log n)^{1/(1-\gamma)}\right)$$

in this case. The next theorem shows that this is also the speed of pointwise convergence of the Robbins–Monro process. Hence the stochastic part "dies out" when $\gamma > 1$.

8.2. Theorem.

Under assumption 8.1 the RM-process given by (25) satisfies

$$\lim_n (\log n)^{1/(1-\gamma)} X_n \qquad \text{converges a.s.}$$

and the possible limit points are

$$\pm[\alpha(\gamma - 1)]^{1/(1-\gamma)} \qquad \text{a.s.}$$

PROOF: It is clear, that $X_n \to 0$ a.s. We define $r(x)$ by

$$r(x) := \frac{f(x)}{\alpha|x|^{\gamma}\operatorname{sgn}(x)} - 1$$

Let $U_n = (\log n)^{1/(\gamma-1)} X_n$. Since

$$\left(\frac{\log(n+1)}{\log n}\right)^{1/(\gamma-1)} = 1 + \frac{1}{(\gamma-1)n\log n}(1 + o(1))$$

U_n satisfies a recursion of the form

$$U_{n+1} = U_n - b_n\left[a\alpha|U_n|^{\gamma}\operatorname{sign}(U_n)(1 + R_n) - \frac{U_n}{\gamma - 1}\right] + c_n W_n$$

where

$$
\begin{aligned}
R_n &= r(X_n) \\
b_n &= \frac{1}{n \log n}(1 + o(1)) \\
c_n &= a\frac{(\log n)^{\frac{1}{(\gamma-1)}}}{n}(1 + o(1)).
\end{aligned}
$$

We choose an $\epsilon > 0$ and an $\eta > 0$ such that $|x| \leq \eta$ implies $|r(x)| \leq \epsilon$. Since $X_n \to 0$ a.s. there is a $N = N(\epsilon)$ such that

$$
P\{\sup_{n \geq N} |X_n| > \eta\} \leq \epsilon. \tag{26}
$$

Define the stopping time τ by

$$
\tau := \inf\{n \geq N : |X_n| > \eta\}.
$$

By (26) the process $V_n := U_{n \wedge \tau}$ coincides with U_n on a set of probability larger than $1 - \epsilon$. Hence it is sufficient to consider the asymptotic distribution of V_n.

Let $h_n(u) := a\alpha|u|^\gamma \text{sgn}(u)(1 + R_{n \wedge \tau}) - \frac{u}{\gamma-1}$. This (random) function has exactly three roots in u, namely

$$
0 \text{ and } \pm (a\alpha(1 + R_{n \wedge \tau})(\gamma - 1)^{1/(1-\gamma)}.
$$

Since $|R_{n \wedge \tau}| \leq \epsilon$ a.s. the nonzero roots lie with probability 1 in the intervals I_1, I_2 where

$$
\begin{aligned}
I_1 &= [(a\alpha(1 + \epsilon)(\gamma - 1))^{1/(1-\gamma)}, (a\alpha(1 - \epsilon)(\gamma - 1))^{1/(1-\gamma)}] \\
I_2 &= -I_1.
\end{aligned}
$$

Let $\varphi(\cdot)$ be a nonnegative twice boundedly differentiable function satisfying

$$
\begin{aligned}
\varphi(u) &= 0 & u \in I_1 \cup I_2 \\
\varphi(u) &= \varphi(-u) \\
\text{sgn}(\varphi'(u) \cdot h_n(u)) &> 0 \\
\varphi(u) &\to \infty & \text{as } |u| \to \infty.
\end{aligned}
$$

Then by Taylor expansion we get on the set $\{\tau > n\}$

$$
\begin{aligned}
\varphi(V_{n+1}) &= \varphi(V_n) - \varphi'(V_n)(b_n h_n(V_n) + c_n W_n) + \\
&\quad + \frac{1}{2}\varphi''(\tilde{V}_n)(b_n h_n(V_n) + c_n W_n)^2
\end{aligned}
$$

where $\tilde{V}_n$ is an intermediate point. Hence denoting by $\mathcal{F}_n$ the sigma field generated by $(V_1, \ldots, V_n)$ we obtain that on the set $\{\tau > n\}$

$$
E(\varphi(V_{n+1})|\mathcal{F}_n) = \varphi(V_n) - b_n E(\varphi'(V_n)h_n(V_n)) + Z_n
$$

with

$$|Z_n| \leq \sup_u |\varphi''(u)|[b_n^2 h_n^2(V_n) + \sigma^2 \cdot c_n^2].$$

Since on $\{\tau = \infty\}$, $|V_n| \leq \epsilon(\log n)^{1/(\gamma-1)}$ by construction we find that $h_n(V_n) = O((\log n)^{1/(\gamma-1)})$ a.s. for large n. Hence $\sum Z_n$ converges a.s.

Since $E(\varphi'(V_n)h_n(V_n)) \geq 0$ with equality only if $V_n \in I_1 \cup I_2 \cup \{0\}$ the well known Robbins-Siegmund Lemma (1971) implies the a.s. convergence of $\varphi(V_n)$ and since $\Sigma b_n = \infty$ the convergence of V_n to the set $I_1 \cup I_n \cup \{0\}$ on $\{\tau = \infty\}$.

Since $h_n'(0) < 0$ the point 0 is not a stable point and can be excluded as limiting point by Theorem 7.16. The cluster points of V_n on $\{\tau = \infty\}$ can only be contained in I_1 or (exclusively) in I_2. Since the length of the intervals I_1 and I_2 was arbitrarily small it follows in fact that U_n converges a.s. and the limit is one of the points

$$\pm(a\alpha(\gamma - 1)^{\frac{1}{(1-\gamma)}}).$$

$\square$

For the cases $\gamma \leq 1$ the following Lemma is crucial, which is related to Lemma 7.9 (cf. Kersting (1978)).

8.3. Lemma. Let V_n fulfill the recursion

$$V_{n+1} = V_n - b_n^2 \cdot h(V_n) + b_n \cdot s(V_n) \cdot W_n$$

where

(a) $\mathbb{E}(W_n) = 0$, $\text{Var }(W_n) = 1, \mathbb{E}(|W_n^3|) \leq c < \infty$

(b) $b_n \geq 0$, $\sum b_n^2 = \infty$, $\sum b_n^3 < \infty$

(c) $h(\cdot)$ is differentiable with $0 < \alpha_1 \leq h'(\cdot) \leq \alpha_2$

(d) $s(\cdot)$ is bounded and Lipschitz.

Then V_n converges in distribution to V, where V has density

$$g(v) = \frac{\text{const.}}{s^2(v)} \exp\left(-2 \int_0^v \frac{h(u)}{s^2(u)} du\right). \tag{27}$$

SKETCH OF THE PROOF: Let ψ be a C^∞-function, which has bounded derivatives of any order. By Taylor expansion,

$$\begin{aligned}\psi(V_{n+1}) &= \psi(V_n) \\ &- \psi'(V_n)(b_n^2 h(V_n) + b_n s(V_n) \cdot W_n)\end{aligned}$$

$$+ \quad \frac{1}{2}\psi''(V_n)(b_n^2 h(V_n) + b_n s(V_n) \cdot W_n)^2$$

$$- \quad \frac{1}{6}\psi'''(\tilde{V}_n)(b_n^2 h(V_n) + b_n s(V_n) \cdot W_n)^3$$

where $\tilde{V}_n$ is an intermediate point. Taking expectations on both sides we get

$$
\begin{aligned}
\mathbb{E}(\psi(V_{n+1})) &= \mathbb{E}(\psi(V_n)) - b_n^2 \mathbb{E}(\psi'(V_n)h(V_n)) + \\
&\quad + \frac{1}{2}b_n^2 \mathbb{E}(\psi''(V_n) \cdot s^2(V_n)) + O(b_n^3) = \\
&= \mathbb{E}\left([(I + b_n^2 L)\psi](V_n) \right) + O(b_n^3)
\end{aligned}
\tag{28}
$$

where I is the identity and L is the linear differential operator

$$(L\psi)(x) = -\psi'(x)h(x) + \frac{1}{2}\psi''(x)s^2(x).$$

This operator has an adjoint L^* in $L^2(\mathbb{R})$, namely

$$(L^*g)(x) = (g(x) \cdot h(x))' + \frac{1}{2}(s^2(x) \cdot g(x))'',$$

in the sense that $\int (L\psi) \cdot g \, dx = \int \psi \cdot (L^*g) \, dx$. It is easily seen that $L^*g \equiv 0$ for the density (27). A contraction argument, which uses (c) shows that the asymptotic distribution of (V_n) is independent of the distribution of V_1. Consequently, one may assume w. l. o. g. that the recursion starts with V_N having a specific distribution. If V_N has density g, then

$$\mathbb{E}((L\psi)(V_N)) = \int (L\psi)(x)g(x)dx = \int \psi(x) \cdot (L^*g)(x)dx = 0.$$

If $n \geq N$, then by (28)

$$
\begin{aligned}
\mathbb{E}(\psi(V_n)) &= \mathbb{E}\left([(I + b_n^2 L)(I + b_{n-1}^2 L) \cdots (I + b_N^2 L)\psi] (V_N) \right) + \sum_{j=N+1}^{n} O(b_j^3) \\
&= \mathbb{E}(\psi(V_N)) + \sum_{j=N+1}^{n} O(b_j^3).
\end{aligned}
$$

Since $\sum_{j=N+1}^{\infty} b_j^3$ is arbitrarily small, if N is large enough, we see that $\lim_{n\to\infty} \mathbb{E}(\psi(V_n))$ must be equal to $\mathbb{E}(\psi(V_N))$ and since the expectations $\mathbb{E}(\psi(\cdot))$ determine the distribution, the limiting distribution of (V_n) must be that of V_N.

Remark, that the limiting distribution (27) is of Gibbs type (see Definition 9), if $s(x)$ is constant. As an application of Lemma 8.3 we consider the recursion

$$X_{n+1} = X_n - \frac{a}{n} f(X_n) - \frac{a}{n} W_n \qquad (29)$$

with

$$f(x) = \alpha \cdot x(1 + r(x)); \qquad |r(x)| = o(|x|).$$

8.4. Theorem. If (X_n) follows the recursion (29) with $2a\alpha > 1$, then $\sqrt{n}X_n$ is asymptotically $N(0, \frac{a^2\sigma^2}{2a\alpha-1})$ distributed.

SKETCH OF THE PROOF: Let $U_n := \sqrt{n}X_n$ and

$$r(x) = f(x) - \alpha x.$$

Then U_n follows the recursion

$$
\begin{aligned}
U_{n+1} &= \sqrt{\frac{n+1}{n}} \left(U_n - \frac{a\alpha}{n} U_n - \frac{a}{n} r\left(\frac{U_n}{\sqrt{n}}\right) - \frac{a}{\sqrt{n}} W_n \right) = \\
&= U_n - \frac{1}{n}\left(a\alpha - \frac{1}{2}\right) U_n - \frac{a}{\sqrt{n}}\sqrt{\frac{n+1}{n}} W_n + \sqrt{\frac{n+1}{n}} \frac{a}{n} r\left(\frac{U_n}{\sqrt{n}}\right).
\end{aligned}
$$

One shows that this recursion has the same asymptotic distribution as the recursion

$$V_{n+1} = V_n - \frac{1}{n}\left(a\alpha - \frac{1}{2}\right) V_n - \frac{a}{\sqrt{n}} \sigma \cdot W'_n$$

where Var $(W'_n) = 1$. An application of Theorem 8.3 with $h(v) = (a\alpha - \frac{1}{2})v$; $s(v) = a \cdot \sigma$ yields the asymptotic distribution with density

$$g(v) = \frac{\text{const.}}{a^2\sigma^2} \exp\left(-\frac{(2a\alpha - 1)v^2}{2a^2\sigma^2}\right)$$

which is a $N\left(0, \frac{a^2\sigma^2}{2a\alpha-1}\right)$-distribution.

□

Major and Revesz (1973) have considered the cases $2a\alpha = 1$ and $2a\alpha < 1$. By the same approach as above one can analyse the case

$$f(x) = \begin{cases} \alpha_1 \cdot x(1 + r(x)) & x \geq 0 \\ \alpha_2 \cdot x(1 + r(x) & x < 0 \end{cases}$$

with $2a\alpha_1, 2a\alpha_2 > 1$. The limiting density is const. $\cdot \exp(-F(x))$, where

$$F(x) = \begin{cases} \frac{(2a\alpha_1-1)x^2}{2a^2\sigma^2} & x \geq 0 \\ \frac{(2a\alpha_2-1)x^2}{2a^2\sigma^2} & x < 0. \end{cases}$$

Another application deals with the case where

$$\lim_{x \to 0} \frac{f(x)}{|x|^\gamma \ \mathrm{sign}(x)} = \alpha$$

for $\frac{1}{2} < \gamma < 1$. By the very same technique we may show that

$n^{\frac{1}{1+\gamma}} X_n$ has an asymptotic distribution

with density const. $\exp\left(-\dfrac{2\alpha}{a\sigma^2} \dfrac{|x|^{1+\gamma}}{1+\gamma}\right).$

The method of Theorem 8.2 extends also to the more dimensional case. We treat only the "linear" case:

8.5. Theorem. The asymptotic distribution of the recursion

$$V_{n+1} = V_n - b_n^2 \, B \cdot V_n - b_n \, W_n$$

where $\mathbb{E}(W_n) = 0$, $\mathrm{Cov}\ (W_n) = C, \sum b_n^2 = \infty, \sum b_n^3 < \infty$, and all eigenvalues of B are positive, is a multivariate $N(0, \Sigma)$-distribution, where Σ is the solution of

$$B\Sigma + \Sigma B^t = C. \tag{30}$$

SKETCH OF THE PROOF: As in the proof of Lemma 8.3 we show that

$$\mathbb{E}(\psi(V_{n+1})) = \mathbb{E}([I + b_n^2 L]\psi](V_n)) + O(b_n^3)$$

for the linear differential operator L

$$L\psi = -(\nabla\psi(v))^t \cdot B \cdot v + \frac{1}{2}tr([\nabla^2\psi(v)] \cdot C).$$

This operator has an adjoint L^*

$$L^*\psi = tr(B \cdot \nabla(\psi \cdot v)) + \frac{1}{2}tr([\nabla^2\psi] \cdot C).$$

The density of the $N(0, \Sigma)$-distribution is

$$g(v) = \ \mathrm{const.} \cdot \exp(-\frac{1}{2}v^t\Sigma^{-1}v).$$

Notice that

$$\begin{aligned}
\nabla g(v) &= -v^t\Sigma^{-1}g(v) \\
\nabla(v \cdot g(v)) &= (-vv^t\Sigma^{-1} + I)g(v) \\
\nabla^2 g(v) &= -\Sigma^{-1}(-vv^t\Sigma^{-1} + I)g(v)
\end{aligned}$$

(I is the identity matrix). The equation $L^*(g) = 0$ follows from

$$tr(B \cdot \nabla(v \cdot g(v))) = \qquad\qquad (31)$$

$$
\begin{aligned}
&= tr(B(-vv^t\Sigma^{-1} + I)g(v)) = tr(B^t(-vv^t\Sigma^{-1} + I)g(v)) \\
&= tr(\Sigma^{-1}\Sigma B^t(-vv^t\Sigma^{-1} + I)g(v)) = \\
&= \frac{1}{2}tr(B\Sigma\Sigma^{-1}(-vv^t\Sigma^{-1} + I)g(v)) + \frac{1}{2}tr(\Sigma B^t\Sigma^{-1}(-vv^t\Sigma^{-1} + I)g(v)) \\
&= \frac{1}{2}tr(C\Sigma^{-1}(-vv^t\Sigma^{-1} + I)g(v)) = -\frac{1}{2}tr(\nabla^2 g(v) \cdot C).
\end{aligned}
$$

We used here the symmetry of Σ and vv^t. The rest of the proof goes along the lines of that of Lemma 8.3. □

The solution of (30) can be made explicit: Using the vec-operation, which makes a $[n \cdot m \times 1]$–vector out of a $[n \times m]$–matrix by putting the colums on top of each other and using the identity

$$\text{vec } (A \cdot B \cdot C) = (C^t \otimes A) \text{ vec } B$$

(Graham (1981), p.25) with $\otimes$ denoting the Kronecker product, we may write

$$\text{vec } \Sigma = [(I \otimes B) + (B \otimes I)]^{-1} \text{ vec } (C)$$

with I being the unit matrix. Another explicit expression is

$$\Sigma = \int_0^\infty \exp(-uB) \cdot C \cdot \exp(-uB^t)du$$

(compare Walk(1977)). For the convergence of this expression it is required that spec $(B) \subseteq (0, \infty)$, where spec is the set of all eigenvalues.

8.6. Corollary. (Fabian(1968)). If X_n is the RM process

$$X_{n+1} = X_n - \frac{a}{n}f(X_n) - \frac{a}{n}W_n,$$

where $\mathbb{E}(W_n) = 0$, Cov $(W_n) \to C, f(x) = A \cdot x + o(x)$, as $x \to x^*$, with spec $(aA) \subseteq (1/2, \infty)$, then $\sqrt{n}X_n$ is asymptotically $N(0, \Sigma)$-distributed, where Σ is the solution of

$$(aA - \frac{1}{2}I)\Sigma + \Sigma(aA^t - \frac{1}{2}I) = C. \qquad\qquad (32)$$

IDEA OF THE PROOF: Denote $U_n = \sqrt{n}X_n$. Show that U_n has the same asymptotic distribution as

$$V_{n+1} = V_n - \frac{1}{n}(aA - \frac{1}{2}I)V_n - \frac{a}{\sqrt{n}}W_n'$$

with $\mathbb{E}(W_n') = 0$, Cov$(W_n) = C$ and apply Theorem 8.5. □

§9 Stopping times

Let

$$X_{n+1} = X_n - a_n f(X_n) - a_n W_n \qquad (33)$$

be a Robbins–Monro–type procedure. In the previous chapters many results concerning the convergence properties of (33) were collected. In practice, however, the procedure must be stopped in finite time. In that case the quality of the approximative solution may be expressed in terms of a confidence region, i.e. a set dependent on the past observations which covers the unknown solution x^* with some predetermined probability.

For getting a confidence region one may either

(i) stop after a predetermined number k of steps. (then the confidence region will have a data–dependent size)

or

(ii) fix the size of the confidence region first and find a stopping time which guarantees that the stopped process lies in a neighborhood of that size of the true solution.

Mathematically, a stopping time τ is an integer valued random variable, such that the set $\{\tau = k\}$ is in the σ-algebra generated by $X_1, \ldots, X_k$. The stopped process X_τ is

$$X_\tau = X_k \qquad \text{on the set} \qquad \{\tau = k\}.$$

If ε is fixed in advance and α is a confidence level, then τ_ε defines a *fixed size confidence region* if

$$\mathbb{P}\{\|X_{\tau_\varepsilon} - x^*\| \le \varepsilon\} \ge 1 - \alpha.$$

Exact level α confidence regions are difficult to obtain, even in the much simpler case of sequential estimation of a mean value (see e.g. Chow and Robbins (1965)). Therefore one is interested in asymptotic level α confidence regions: By this a family of stopping times (τ_ε) indexed by ε is meant such that

$$\lim_{\varepsilon \to 0} \mathbb{P}\{\|X_{\tau_\varepsilon} - x^*\| \le \varepsilon\} \ge 1 - \alpha. \qquad (34)$$

The speed of convergence in the limit (34) may heavily depend on the starting value X_1. An asymptotic level α confidence region is called *uniform in the starting value*, if

$$\lim_{\varepsilon \to 0} \sup_x \mathbb{P}\{\|X_{\tau_\varepsilon} - x^*\| \le \varepsilon | X_1 = x\} \ge 1 - \alpha.$$

There are some stopping procedures defined in the literature. Farrell (1962) considers only the univariate case and defines a stopping time in the following way: He starts two different approximations at a very small and a very large value respectively and stops at the first moment the two procedures overlap. His procedure is uniform in the starting value, but covers only the univariate case.

Sielken (1973) derives a stopping time from the asymptotic distribution: It is known that with $a_n = \frac{a}{n}$, Cov $(W_n) \to C$ and $f(x) = A \cdot x + o(x)$ as $x \to x^*$, $\sqrt{n}X_n$ is asymptotically $N(0, \Sigma)$ distributed, where Σ is the solution of (32) (see Corollary 8.6). Suppose that $\hat{A}_n$ and $\hat{C}_n$ are estimators of A resp. C. Then an estimate $\hat{\Sigma}_n$ for Σ can be found by inserting the estimates for A and C in (32). Now the stopping time τ_ε is the smallest n so that

$$\mathbb{P}\{\|\frac{1}{\sqrt{n}}X\|^2 \leq \varepsilon\} \geq 1 - \alpha,$$

where X is distributed according to $N(0, \hat{\Sigma})$. This stopping rule is not uniform in the starting value and of limited practical interest because its performance is terribly bad in deterministic situations: In fact, if the covariance matrix equals zero then the procedure stops immediately at the starting value.

Another stopping rule was given by Stroup and Braun (1982). Their rule is based on a χ^2 criterion and has the disadvantage not to stop at all if the noise is zero.

Pflug (1990) considers the case of fixed stepsizes and designes a test for detecting whether the approximation has reached its stationary behavior and oscillates near the solution. For an overview of practically used stepsize rules and stopping times see Pflug (1988).

§10   Applications of stochastic approximation methods

Stochastic Approximation is a group of methods for finding solutions of equations

$$(E) \left\|\begin{array}{l} f(x) = 0 \\ x \in S \end{array}\right.$$

or optimization problems

$$(P) \left\|\begin{array}{l} F(x) = min! \\ x \in S \end{array}\right.$$

in cases where $f(x)$ resp. $F(x)$ are only observable together with some random noise. Notice that (P) is considered as a special case of (E), if f is the gradient of F and random observations of the gradient are possible.

The competitors of stochastic approximation methods are the so called *response surface methods*. The distinguishing factor between these two groups of methods is the amount of information about F stored from step to step and the way of approximating this function by a member of some simpler class.

- The Robbins–Monro procedure stores only the current approximation value X_n and approximates F locally by a linear function via its gradient f.

- The Kiefer–Wolfowitz procedure also stores only the current value X_n and approximates F linearly, by taking 2^k additional design points and a linear interpolation.

- The Polyak–Ruppert modification of the Robbins–Monro procedure stores X_n and the arithmetic mean $\overline{X}_n$ of all previous design points. It was independently invented by Polyak and Ruppert.

- In general, *adaptive design methods* select the next design point X_{n+1} according to the information gained so far. They are in contrast to the *fixed design methods*, where the design points used to approximate the unknown function F are fixed in advance. The main advantage of adaptive design methods lies in the fact that more design points are generated in the area of interest near the solution and less effort is done to investigate the behavior of F in regions of no interest. The theory of experimental design (Fedorov (1972)) solves the problem of optimally placing a new design point given all information gathered so far. We speak of a *response surface* method, if a curve of some parametric class is fitted first to the design points and the new design point is then found by experimental design technique. Typically *linear*, *quadratic* or *cubic* functions are fitted. The method is called *local*, if only the design points in the neighborhood of the presumptive solution are used for the approximation. Otherwise the method is called *global*. See Marti (1992) for a review on response surface methods.

The typical application of stochastic approximation is characterized by an interplay between data collection and the calculation of the new design point X_{n+1}.

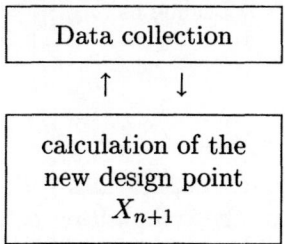

Case of *design-dependent random observations*

Sometimes however, the random observations do not depend on the design points and could be collected in advance.

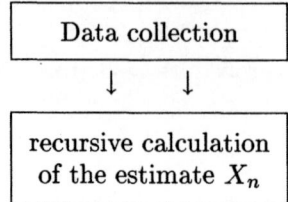

Case of *design-independent random observations*

Another distinction can be made according to the way of generating the random variables:

real-life experiment versus *computer simulation.*

We may thus categorize the application areas into four groups:

distribution of observation	*data source*	
	real-life data	computer simulation
design-dependent	"classical situation"	stoch. optimization with decision-dependent probability laws
design-independent	recursive statistical estimation Pattern recognition and learning schemes	stoch. optimization with decision-independent probability laws

10.1. The classical situation

In biological, medical or technical applications x is an experimental parameter, such as the dose of a drug or the composition of an alloy and the goal is to find the parameter value x which either minimizes an objective function F or solves an equation $f(x) = 0$.

Example. Let x denote the dose of a poisonous drug and

$$G(x) = \text{ probability that a rat dies from dose } x.$$

If we expose a rat to the drug x we may observe the random variable

$$Y_x = \begin{cases} 1 & \text{rat dies from dose } x \\ -1 & \text{rat survives.} \end{cases}$$

Let $f(X_n) = \mathbb{E}(Y_{X_n}) = G(X_n) - (1 - G(X_n)) = 2 \cdot G(X_n) - 1$. The root of f corresponds to the letal dose 50 ($LD50$), which is the dose having a survive probability of 0.5. With the recursion

$$X_{n+1} = X_n - a_n Y_{X_n}$$

one may therefore estimate the $LD50$ by sequentially adjusting the dose and making one experiment at each step.

10.2. Statistical applications

Let $\xi_1, \xi_2, \ldots, \xi_k$ be a sequence of i.i.d. observations stemming from a distribution G_γ, where γ is some possibly infinite dimensional parameter. We want to estimate a finite dimensional parameter $\vartheta = \vartheta(\gamma)$ from (ξ_i). Suppose that there is a function $Y(\xi, \vartheta)$ such that

$$\mathbb{E}_\gamma(Y(\xi, \vartheta(\gamma)) \equiv 0. \tag{35}$$

Then, if some regularity assumptions are fulfilled, we may estimate the unknown ϑ by the recursion

$$X_{n+1} = X_n - a_n Y(\xi_n, X_n).$$

Example. Estimation of Quantiles.
Let the unknown parameter ϑ be the α-quantile of G_γ:

$$G_\gamma(\vartheta(\gamma)) \equiv \alpha .$$

Then the function $Y(\xi, \vartheta) := \mathbf{1}_{\{\xi \le \vartheta\}} - \alpha$ fulfills (35) and we may estimate ϑ by

$$X_{n+1} = X_n - a_n(\mathbf{1}_{\{\xi_n \le X_n\}} - \alpha)$$

This estimate is recursive, in contrast to the sample α-quantile.

Example. Recursive estimation of general parameters:
Let $(g_\vartheta)_{\vartheta \in I\!\!R^k}$ be a family of densities w.r.t. a σ-finite measure ν indexed with a real parameter vector ϑ. We assume that $\vartheta \mapsto \sqrt{g_\vartheta}$ is L_2-differentiable, i.e. there is a vector of L_2-functions h_ϑ such that

$$\left\| \sqrt{g_{\vartheta'}} - \sqrt{g_\vartheta} - (\vartheta' - \vartheta)^t h_\vartheta \right\|_{L_2} = o(\|\vartheta' - \vartheta\|)$$

as $\vartheta' \to \vartheta$. Define

$$Y(\xi, \vartheta) := \frac{h_\vartheta(\xi)}{\sqrt{g_\vartheta(\xi)}}.$$

If ξ has density g_{ϑ_0}, then

$$\mathbb{E}(Y(\xi, \vartheta_0)) = \int h_{\vartheta_0} \cdot \sqrt{g_{\vartheta_0}} \, d\nu = 0,$$

i.e. Y fulfills (35). An efficient estimate of ϑ is given by the recursion

$$X_{n+1} = X_n - a_n \cdot I^{-1}(X_n) \cdot \frac{h_{X_n}}{\sqrt{g_{X_n}}} \qquad (36)$$

where $I(\vartheta)$ is the Fisher-information matrix

$$I(\vartheta) = 4 \int h_\vartheta \cdot h_\vartheta^t d\nu.$$

Conditions for the convergence of (36) were studied by Fabian(1978).

10.3. Pattern recognition and learning schemes

A system is able to learn if it may change its own behavior according to outer influences. One may distinguish between *incremental learning*, where each new information is added to the internal memory and *adaptive learning*, where the size of the internally stored information remains unchanged and only its content is adapted. If, in the latter case, the internally stored information is just a real parameter vector, we speak about *parametric learning*.

Stochastic approximation is the prototype of parametric learning. We illustrate this by an example from pattern recognition:

Example. Learning the optimal linear discrimination function

Statistical Pattern Recognition is the decision about a group membership on the basis of the observation of a random variable ξ. Suppose for simplicity that there are only two possible groups:

group 1 has a relative frequency π_1 and ξ has d.f. G_1 in this group
group 2 has a relative frequency π_2 and ξ has d.f. G_2 in this group.

For learning the best linear discrimination rule between the groups one must observe ξ together with the group number. The goal of the learning phase is to set the parameters w (a k-vector) and w_0 (a scalar) such that the linear decision rule:

$$\text{decide for group} \begin{cases} 1 & \text{if} \quad \xi^t \cdot w \geq w_0 \\ 2 & \text{if} \quad \xi^t \cdot w < w_0 \end{cases}$$

is optimal. By introducing the augmented variables

$$\eta_i = \begin{cases} \binom{-1}{\xi_i} & \text{if the observation is from group 1} \\[2mm] \binom{1}{-\xi_i} & \text{if the observation is from group 2} \end{cases}$$

and the decision rule vector $x = \binom{w_0}{w}$, the decision is correct iff $\eta_i^t \cdot x \geq 0$. The quality (and therefore the optimality) of a decision is measured by some objective function. Here are typical examples:

(i) "Probability of misclassification"

$$F_1(x) = \mathbb{P}\{\eta^t \cdot x < 0\} = \min!$$

(ii) "Rosenblatt's perception rule"

$$F_2(x) = \mathbb{E}(-\eta^t x \cdot \mathbf{1}_{\{\eta^t x < 0\}}) = \min!$$

(iii) "Quadratic error function"

$$F_3(x) = \mathbb{E}((\eta^t \cdot x - 1)^2) = \min!$$

For all three criterion functions there is an unbiased estimate available, namely

(i) $\mathbf{1}_{\{\eta^t \cdot x < 0\}}$

(ii) $-\eta^t \cdot x \cdot \mathbf{1}_{\{\eta^t \cdot x < 0\}}$

(iii) $(\eta^t \cdot x - 1)^2$

and one may use a Kiefer-Wolfowitz-type procedure to find the optimal x^*. As the Robbins-Monro procedure has a higher rate of convergence compared to the Kiefer-Wolfowitz procedure, it is of importance to find a *stochastic gradient* of F, i.e. random functions $h_j(x, \omega)$ with expectation

$$\mathbb{E}(h_j(x, \cdot)) = \frac{\partial}{\partial x_j} F(x). \tag{37}$$

The notion of L_1-derivative is of importance here.

10.4. Definition. Let $\{H(x, \omega)\}_{x \in I\!\!R^k}$ be a family of L_1-functions. A vector of L_1-functions $h(x, \omega) = (h_1(x, \omega), \ldots, h_k(x, \omega))$ is called the L_1-derivative of H at x, if

$$\|H(y, \cdot) - H(x, \cdot) - (y - x)^t h(x, \cdot)\|_{L_1} = o(\|y - x\|)$$

as $y \to x$.

An immediate consequence of this definition is that if $F(x) = \mathbb{E}(H(x, \cdot))$, and h is the L_1-derivative of H, then h is also a stochastic gradient of F in the sense of (37).

The function F_1 from above does not possess a L_1-derivative, whereas the L_1-derivatives of F_2 and F_3 are

(ii) $-\eta^t \cdot \mathbf{1}_{\{\eta^t \cdot x < 0\}}$

(iii) $2\eta^t \cdot (\eta^t \cdot x - 1)$.

A learning scheme of the RM–type for the optimal x w.r.t. F_2 is for example

$$X_{n+1} = X_n + a_n \cdot \eta_i \mathbf{1}_{\{\eta_i^t X_n < 0\}}.$$

10.5. Stochastic optimization with decision–independent probabilities

The general structure of such a problem is

$$\left\| \begin{array}{l} F(x) = \int H(x, \omega)\, d\mu(\omega) = \min! \\ x \in S \end{array} \right.$$

If a stochastic gradient of $h(x, \omega)$ of $F(\cdot)$ in the sense of (37) exists, this problem may be solved by the constrained RM algorithm

$$X_{n+1} = \pi_S(X_n - a_n h(X_n, \cdot)) \tag{38}$$

where π_S is the projection onto S which is assumed to be convex and closed.

Example. Aircraft Allocation. An airline company has to decide about the number of aircrafts of type i to fly on route j. The number of passengers on route j is random. Introduce the following quantities

$x_{ij} \cdots$ number of aircrafts of type i on route j (decision variable)
$t_i \ \cdots$ passenger capacity of aircraft type i
$b_i \ \cdots$ number of aircrafts of type i available
$c_{ij} \cdots$ cost of one aircraft of type i on route j
$q_j \ \cdots$ loss per rejected passenger
$\xi_j \ \cdots$ number of passengers on route j (random variable)

The mathematical problem formulation is

$$\left\| \begin{array}{l} \sum_{ij} c_{ij} x_{ij} + \mathbb{E}\left(\sum_j q_j (\xi_j - \sum_i x_{ij} t_i)^+\right) = \min! \\ \sum_j x_{ij} \leq b_i \\ x_{ij} \geq 0. \end{array} \right. \tag{39}$$

Here $x^+ = \max(x, 0)$. The set of constraints of (39) is a compact polyhedron. A stochastic gradient of the so called "recourse function"

$$\mathbb{E}(\sum_j q_j (\xi_i - \sum_i x_{ij} t_i)^+)$$

w.r.t. x_{ij} is

$$-t_i \cdot q_j \cdot \mathbf{1}_{\{\xi_j - \sum_i x_{ij} t_i \geq 0\}} \cdot$$

By sampling independent realizations of the random variables $\xi_j^{(1)}, \xi_j^{(2)}, \ldots$ for each step of the RM-algorithms we may solve (39) by the stochastic approximation algorithm (38).

10.6. Stochastic optimization with decision–dependent probabilities

The general form of such a problem is

$$\left\|\begin{array}{l} F(x) = \int H(x,\omega)\,d\mu_x(\omega) = \min! \\ x \in S \end{array}\right.$$

with μ_x a probability measure depending on the decision variable x. Typically, μ_x is the stationary distribution of a stochastic process, which is controlled by the parameter x.

Example. An open queuing network (Jackson Network) consists of k service nodes. Customers arrive at the i-th node with intensity λ_i (Poisson instream) and branch after service at node i to node j with given probability r_{ij}. With probability $1 - \sum_j r_{ij}$ they leave the system. λ_i and r_{ij} are given constants. The decision vector is the vector of service intensities $x = (x_1, \ldots, x_k)$ at the respective nodes (exponential service times). The goal is to minimize the total mean waiting time of the customers under the budget restriction

$$\sum x_i \le c.$$

Here, μ_x is the stationary distribution of the sums of the waiting times and H is the identy. The queuing process may easily be simulated by computer simulation to get good estimates of the objective function (Remark that the long-run behavior of an ergodic stochastic system is only an approximation of its stationary behavior). Thus one may apply a KW–type procedure. The most important issue however is to find estimates of $\frac{\partial}{\partial x} F(x)$ in order to make the RM-procedure applicable (see next section).

10.7. Stochastic gradients

Let $\{\mu_x\}_{x \in \mathbb{R}}$ be a family of probability measures on the Borel σ-algebra of some metric space R. For simplicity we assume that x is a one–dimensional real parameter. Suppose that there is a σ-finite measure ν, which dominates all μ_x and let $\phi_x(\omega)$ be the density

$$\phi_x(\omega) := \frac{d\mu_x}{d\nu}(\omega).$$

There are well known methods to generate a random variable W_x with a given density ϕ_x on a computer. Let H be a bounded continuous function. Then $H(W_x)$ is an unbiased estimate of $F(x) = \int H(\omega)\phi_x(\omega)\,d\nu(\omega)$. The purpose of this section is to show how stochastic gradients, i.e. unbiased estimates of $\frac{\partial}{\partial x} F(x)$ may be generated.

Suppose that $x \mapsto \phi_x$ is L^1-differentiable in the sense of Definition 10.4 with derivative $\phi'_x(\cdot)$. Since ϕ'_x is not a probability density, we cannot directly sample from it. However, we may find another probability measure γ with density $\psi(\cdot) = \frac{d\gamma}{d\nu}$ having the property that

$$\{w : \psi(w) = 0\} \subseteq \{w : \phi'_x(w) = 0\} \quad \nu - \text{ a.e.} \tag{40}$$

and sample a random variable V with density ψ. Then $\frac{\phi'_x(w)}{\psi(w)}$ is ν-a.e. well defined and for bounded H

$$H(V) \cdot \frac{\phi'_x(V)}{\psi(V)} \text{ is an unbiased estimate of } \frac{\partial}{\partial x} F(x),$$

since

$$\mathbb{E}(H(V) \cdot \frac{\phi'_x(V)}{\psi(V)}) = \int H(w) \frac{\phi'_x(w)}{\psi(w)} \psi(w) d\nu(w) = \frac{\partial}{\partial x} \int H(w) \phi_x(w) d\nu(w).$$

All probability densities ψ with property (40) give unbiased estimates for $\frac{\partial}{\partial x} F(x)$. The most popular choice for ψ is ϕ_x itself:

10.8. Definition. The function

$$l_x(w) := \frac{\phi'_x(w)}{\phi_x(w)} \tag{41}$$

is called the *score function*, the estimate

$$Y = H(Z_x) \cdot l_x(Z_x),$$

where Z_x has distribution μ_x is called the *score function estimate* of $\frac{\partial}{\partial x} F(x)$ (Rubinstein (1986).

Notice that

$$\mathbb{E}(l_x(Z_x)) = \int \phi'_x(w) d\nu(w) = 0 \tag{42}$$

since $\int \phi'_x(w) d\nu(w) = \frac{\partial}{\partial x} \int \phi_x(w) d\nu(w) = \frac{\partial}{\partial x} 1 = 0$. The choice $\psi = \phi_x$ is in accordance with (40) as may be easily shown.

What is the best choice for the alternative measure γ ? Letting ψ run through all densities with property (40), one may show that the variance of Y is minimized, if $\psi(w)$ is chosen proportional to $H(w) \cdot |\phi'_x(w)|$ (Pflug(1992)).

It is inconvenient that the optimal ψ depends on the function H, which is not known as whole. A universal choice of ψ which is good for nearly constant H is proportional to $|\phi'_x(w)|$. Let $\psi_0(w) := (2c_x)^{-1} \cdot |\phi'_x(w)|$ where

$$c_x = \int \max(\phi'_x(w), 0) \, d\nu(w) = -\int \min(\phi'_x(w), 0) \, d\nu(w)$$

$$= \frac{1}{2} \int |\phi'_x(w)| \, d\nu(w).$$

If V_0 has density ψ_0, then

$$Y_0 := H(V_0) \cdot \text{ sign } (\phi'_x(V_0)) \tag{43}$$

is the corresponding unbiased estimate of $\frac{\partial}{\partial x} F(x)$.

A more general concept than the differentiation of densities is the notion of weak derivative:

10.9. Definition. A triple $(c_x, \dot{\mu}_x, \ddot{\mu}_x)$ consisting of a constant c_x and two probability measures $\dot{\mu}_x$ and $\ddot{\mu}_x$ is called the weak derivative of μ_x w.r.t. x, if for all bounded continuous functions G

$$\lim_{s \to 0} \frac{1}{s} | \int G(\cdot)\, d\mu_{x+s}(\cdot) - \int G(\cdot)\, d\mu_x(\cdot) \quad -$$

$$-c_x \cdot s[\int G(\cdot)\, d\dot{\mu}_x(\cdot) \quad - \quad \int G(\cdot)\, d\ddot{\mu}_x(w)]| = 0. \tag{44}$$

It is easy to see that if the family $\{\mu_x\}$ has L^1-differentiable densities, it has also a weak derivative: We have only to set c_x as before and $\dot{\mu}_x$ resp. $\ddot{\mu}_x$ as the measures with densities

$$\frac{d\dot{\mu}_x}{d\nu} = \frac{1}{c_x} \max(\phi'_x, 0) =: \dot{\phi}_x$$

$$\frac{d\ddot{\mu}_x}{d\nu} = \frac{-1}{c_x} \min(-\phi'_x, 0) =: \ddot{\phi}_x.$$

The converse is not true: the weak derivative may exist in cases where the property of L^1-differentiability does not hold.

If $\dot{Z}$ is distributed according to density $\dot{\phi}_x(w)$ and $\ddot{Z}$ is distributed according to density $\ddot{\phi}_x(w)$, then

$$Y_* := c_x \cdot (H(\dot{Z}) - H(\ddot{Z})) \tag{45}$$

is an unbiased estimate for $\frac{\partial}{\partial x} F(x)$. This *weak derivative estimate* Y_* is similar to Y_0 defined in (43), but has smaller variance in general.

Approximative Derivatives for stationary distributions of Markov Chains.

Let P_x be a Markov transition operator on the metric state space (R, r). We assume that P_x is a regular transition operator, i. e.

(i) $w \mapsto P_x(w, A)$ is measurable for each Borel set A,

(ii) $A \mapsto P_x(w, A)$ is a measure for each w.

Suppose that there is a measure ν on the Borel σ–algebra $\mathcal{A}$ such that P_x has a transition density $p_x(\cdot, \cdot)$

$$P_x(w, A) = \int_A p_x(w, v)\, d\nu(v).$$

If the starting distribution has density ψ_x w.r.t. ν, then a Markov sequence

$$M_x(0) = w_0, M_x(1) = w_1, \cdots, M_x(n) = w_n$$

has density

$$\phi_x(w_0, \cdots, w_n) = \psi_x(w_0) \cdot p_x(w_0, w_1) \cdots p_x(w_{n-1}, w_n)$$

w.r.t. $\nu \otimes \cdots \otimes \nu$ (n+1 times).

Suppose that $x \mapsto p_x(w, \cdot)$ is L^1-differentiable for every w with derivative $p'_x(w, v)$ (compare (41)).

The score function l_x is defined in the Markov case as

$$l_x(w_0, \cdots, w_n) := \frac{\phi'_x(w_0, \cdots, w_n)}{\phi_x(w_0, \cdots, w_n)} = \frac{\psi'_x(w_0)}{\psi_x(w_0)} + \sum_{i=1}^{n} \frac{p'_x(w_{i-1}, w_i)}{p_x(w_{i-1}, w_i)}. \qquad (46)$$

If we replace in (46) the dummy variables w_i by the values of the Markov process $M_x(i)$, we get the stochastic process

$$W(n) := l_x(M_x(0), \cdots, M_x(n)),$$

which is by (42) a zero-mean martingale, which is called the *sensitivity process*.
The score function estimate of $\frac{\partial}{\partial x} \int H(w) \, d\mu_x(w)$ is

$$Y_x^{(s)} := \frac{1}{N} \sum_{i=1}^{N} H(M_x(i)) \cdot W_x(i). \qquad (47)$$

(Rubinstein(1986)). For a detailed discussion on stochastic gradients for Markov Chains (including weak derivatives) see Pflug (1992).

References for Part II:

Chiang T.S., Hwang C.R., Sheu S.J. (1987) Diffusions for global optimization in $\mathbb{R}^n$. SIAM J. Control Optim., Vol. 25, 737 – 752.

Chow Y., Robbins H. (1965). On the asymptotic theory of fixed–width sequential confidence intervals.
Ann. Math. Statist., Vol. 36, 457 – 462.

Chung K.L. (1954). On a stochastic approximation method.
Ann. Math. Statist., Vol. 25, 463 – 483.

David (1970). Order statistics.
J. Wiley and Sons, New York.

Dekker A., Aarts E. (1991). . Global optimization and simulated annealing.
Math. Programming, Vol. 50, 367 – 393.

Dixon L., Szegö G. (1975). Towards Global Optimization.
North Holland, Amsterdam.

Fabian V. (1968). On asymptotic normality in stochastic approximation.
Ann. Statist., Vol. 39, 1327 – 1332.

Fabian V. (1978). On asymptotically efficient recursive estimation.
Ann. Statist., Vol. 6., No 4, 854 – 856.

Farell R.H. (1962). Bounded length confidence intervals for the zero of a regression function.
Ann. Math. Statist., Vol. 33, 237 – 247.

Fedorov V. (1972). Theory of optimal experiment.
Academic press, New York.

Föllmer H. (1988). Random Fields and Diffusion Processes.
Ecole d'été de probabilité de St.-Flour XV - XVII, Springer Lecture Notes 1362.

Gelfand S.B., Mitter S.K. (1991). Recursive stochastic algorithms for global optimization in $\mathbb{R}^d$.
SIAM J. Control, Vol. 29, No. 5, 999 – 1018.

Ge Renpu (1990). A filled function method for finding a global minimizer of a function of several variables. Math. Progr. Vol. 46.

Geman S., Hwang C.R. (1986). Diffusions for global optimization. Siam J. Control and Optimization, Vol.34, No. 3, 1031 – 1036.

Gihman J., Skorohod A. (1968). Stochastic differential equations. Kiev: Nauk. dumka (in Russian)

Graham A. (1981). Kronecker procucts and matrix calculus.
Ellis Horwood.

Heyde C.C. (1974). On martingale limit theory and strong convergence results for stochastic approximation procedures.
Stoch. Proc. and Appl., Vol. 2, 359 – 370.

Högnäs G. (1986). Comparison for some nonlinear autoregressive processes.
J. Time Series Analysis, Vol.7., No.3, pp. 205 – 211.

Hwang C.R. (1980). Laplace's method revisited: weak convergence of probability measures. Ann Probab., Vol.8, 1177 – 1182.

Karlin S., Taylor H. (1981). A second course in stochastic processes.
Academic press, New York.

Kersting G.D. (1977). Some results on the asymptotic behavior of the Robbins-Monro procedure.
Bull. Int. Stat. Inst., Vol. 47, 327 – 335.

Kersting G.D. (1978). A weak convergence theorem with application to the Robbins-Monro process.
Ann. Prob., Vol. 6., No. 6, 1015 – 1025.

Kushner H., Hai-Huang (1981). Asymptotic properties of stochastic approximation with constant coefficients.
SIAM J. Control Vol. 19, 87 – 105.

Kushner H. (1987). Asymptotic global behavior for stochastic approximation and diffusions with slowly decreasing noise effects: global minimization via Monte Carlo.
Siam J. Appl. Math., Vol. 47, No. 1, 169 – 185.

Major P., Revesz P. (1973). A limit theorem for the Robbims–Monro approximation.
Z. Wahrscheinlichkeitstheorie verw. Geb., Vol. 27, 79 – 86.

Marti K. (1980). On Accelerations of the Convergence in Random Search Methods.
Meth. Oper. Res., Vol. 37, 391 – 406.

Marti K. (1992). Semi-Stochastic Approximation by the Response Surface Methodology. Optimization.

Matyas J (1965). Random Optimization. Automation and Remote Control, Vol. 26, 246 – 253.

v.Mises R., Pollaczek-Geiringer H (1929). Praktische Verfahren der Gleichungsauflösung. Z. angew. Math. Mech. Vol. 9, 58 – 77.

Nevel'son M.B., Hasminskij R.S. (1972). Stochastic approximation and recurrent estimation.
Nauka, Moskwa (in Russian). Translated in Amer. Math. Soc. Transl. Monographs, Vol.24, Providence, R.I.

Neveu J. (1974). Discrete Parameter Martingales.
North Holland, Amseterdam.

Pflug G. (1986). Stochastic optimization with constant step–size. Asymptotic laws.
SIAM J. of Control, Vol. 24, No. 4, 655 – 666.

Pflug G. (1988). Stepsize rules, stopping times and their implementation in stochastic quasigradient algorithms.
In: Numerical Techniques for Stoch. Optimization (Y. Ermoliev, R. Wets eds.), Springer Series in Computational Mathematics.

Pflug G. (1989). Sampling derivatives of probability measures.
Computing, Vol. 42, 315 – 328.

Pflug G. (1990). Non–asymptotic Confidence Bounds for Stochastic Approximation Algorithms with Constant Step Size.
Monatsh. Math., Vol. 110, 297 – 314.

Pflug G. (1991). A note on the comparison of stationary laws of Markovian processes.
Statistics and Probability Letters, Vol. 11, No. 4, 331 – 334.

Pflug G. (1992). Gradient estimates for the performance of Markov Chains and Discrete Event Processes.
to appear in: Annals of OR.

Pflug G. Ch., Schachermayer W. (1992). Cofficients of ergodicity for stochastically monotone Markov Chains. to appear in: Advances of Applied Probability.

Polyak B. (1991). Novi metod tipa stochasticekoi approksmacii.
Automatika i Telemechanika No.7, 98 - 107 (in Russian).

Rachev S.T. (1984). The Monge-Kantorovich mass transformation problem and its stochastic applications.
Theory of Probability and its Applications, Vol. 29, No.4, 647 – 676.

Revuz D. (1975). Markov Chains.
North Holland Publ. Comp. Amsterdam.

Robbins H., Monro S. (1971). A stochastic approximation method.
Ann. Math. Statist., Vol. 22, 400 – 407.

Robbins H., Siegmund D. (1971). A convergence theorem for nonnegative almost supermartingales and some applications. Optimizing methods in Statistics, ed. by J.S. Rustagi.
Academic Press, New York, 233 – 257.

Rubinstein R. (1986). The score function approach for sensitivity analysis of computer simulation models.
Mathematics and Computers in Simulation, Vol. 28, 351 – 379.

Sielken R.L. (1973). Some stopping rule for stochastic approximation procedures.
Z. Wahrscheinlichkeitstheorie verw. Geb., Vol. 27, 79 – 86.

Solis F., Wets R. (1981). Minimization by random search techniques.
Mathematics of Operations Research, Vol. 6, No. 1, 19 – 30.

Strassen V. (1965). The existence of probability measures with given martingals.
Ann. Math. Statist., Vol. 36, 423 – 439.

Stroup D.F., Braun H.I. (1982). A new stopping rule for stochastic approximation.
Z. Wahrscheinlichkeitstheorie verw. Geb., Vol. 60, 535 – 554.

III Applications to adaptation algorithms

Lennart Ljung
Department of Electrical Engineering,
Linköping University,
S-581 83 Linköping, Sweden

§ 11 Adaptation and tracking

To cope with a changing world is a basic concern. In many engineering systems there are requirements of adaptability. Examples include adaptive control (adjust the regulator to the current properties of the system), adaptive filtering (continuously tuning a signal processing filter to allow for optimal action at all times), adaptive prediction and so on.

At the heart of all such adaptive mechanisms, there is a feature that tracks the time-varying properties of the underlying system or signal. We shall in this part describe how algorithms for such tracking are derived and analysed.

11.1. Model structures

The first step is to define a structure within which we look for a description of the object to be monitored or controlled. In general this structure takes the form of a non-linear regression

$$y(t) = g(t, \theta, \varphi(t)) + e(t) \tag{1}$$

Here $y(t)$ and $\varphi(t)$ are made up from measured signals that are available at time (sample number) t. We shall often refer to the variable $y(t)$ - which is to be "explained" or predicted by the model - as "the output". The vector $\varphi(t)$ consists of present and past measurements. It is quite conceivable that $\varphi(t)$ also contains past values of $y(t)$, $s < t$. The variable $e(t)$ represents various disturbances that also affect the measurements. Finally, the vector θ contains the parameters that we require to describe the relationsship (1). The objective is thus to estimate this parameter vector based on

$$y(s), \quad \varphi(s) \quad s \leq t \tag{2}$$

We shall denote this estimate by

$$\hat{\theta}(t) \tag{3}$$

A common special case is where (1) is a linear regression:

$$y(t) = \varphi^T(t)\theta + e(t) \tag{4}$$

With

$$\theta = (a_1 \ldots a_n)^T \quad \varphi(t) = (-y(t-1) \ldots y(t-n))^T \tag{5}$$

this gives the familiar AR-representation of a signal:

$$y(t) + a_1 y(t-1) + \ldots + a_n y(t-n) = e(t) \tag{6}$$

With

$$\theta = (b_1 \ldots b_n)^T \quad \varphi(t) = (u(t-1), \ldots, u(t-n))^T \tag{7}$$

we have a so called finite impulse-response (FIR) model for the relationship between the (exogenous) signal $\{u(t)\}$ and the output:

$$y(t) = b_1 u(t-1) + \ldots + b_n u(t-n) \tag{8}$$

Clearly (4) covers many more models, and (1) indeed is sufficient to describe all commonly used models in control and signal processing (if the dimension of $\varphi(t)$ is allowed to increase with t). See e.g. [8] for a comprehensive treatment of this.

We shall in the next section discuss how to derive algorithms for the estimation of θ. Before that, let us however comment on the time-varying character of the system/signal description. It is in the nature of the matter that the "true" parameter description θ indeed varies with time:

$$y(t) = g(t, \theta(t), \varphi(t)) + e(t) \tag{9}$$

or

$$y(t) = \varphi^T(t)\theta(t) + e(t) \tag{10}$$

We shall denote the parameter change by $w(t)$

$$\theta(t) = \theta(t-1) + w(t) \tag{11}$$

and we shall in the next section discuss various assumption about $\{w(t)\}$ and their consequences.

This part of the book will be organized as follows. The following section will deal with various ways of deriving tracking algorithms. In § 13 we briefly quote some convergence and asymptotic distribution results, while the case of estimating the tracking ability is covered in § 14.

§ 12 Algorithm development

12.1. Algorithms for general non-linear regressions

As we noted in § 11 most models for dynamical systems can be cast into the form (1)

$$y(t) = \hat{y}(t \mid \theta) + e(t) \tag{12}$$

where $\hat{y}(t \mid \theta) = g(t, \theta, \varphi(t))$ is a general function of past input-output data and of the parameter vector θ. The notation $\hat{y}$ emphasizes the interpretation of this quantity as a predictor. We note in passing that also multi-layered perceptions (neural networks) are special cases of (12) (θ then corresponds to the weights in the interconnections). If we assume $\{e(t)\}$ in (12) to be a sequence of independent random variables with probability density functions $f(x,t)$ and define

$$\ell(x,t) = -\log f(x,t) \tag{13}$$

the negative log likelihood function for the estimation of θ in (12) will be

$$V(\theta) = \sum_{k=1}^{t} \ell(\varepsilon(k,\theta),k) \tag{14}$$

where

$$\varepsilon(t,\theta) = y(t) - \hat{y}(t \mid \theta) \tag{15}$$

To cope with the fact that the actual parameter value may be time-varying, it should be natural - on an *ad hoc* basis - to give more weight in (14) to more recent measurements than to older ones. This would lead to a weighted prediction error criterion

$$V_t(\theta) = \sum_{k=1}^{t} \beta(t,k)\ell(\varepsilon(k,\theta),k) \tag{16}$$

Here $\ell(\cdot)$ is a scalar valued function that - in some sense - measures the "size" of the prediction error ε

In the off-line case, (16) is typically minimized by iterative search, e.g. of the Gauss-Newton type. A basic approach to adaptation is to perform one iteration for the minimization of (16) at the same time as one more observation (t increased one unit) is obtained. This approach is detailed in [8], Chapter 11. If

$$\beta(t,k) = \prod_{j=k+1}^{t} \lambda(j) \tag{17}$$

the resulting algorithm is of the form

$$\hat{\theta}(t) = \hat{\theta}(t-1) + R^{-1}(t)\psi(t)\ell'_\varepsilon(\varepsilon(t),t) \tag{18}$$

$$R(t) = \lambda(t)R(t-1) + \psi(t)\ell''_{\varepsilon\varepsilon}(\varepsilon(t),t)\psi^T(t) \tag{19}$$

(See (11.52) of [8]). Here $\psi(t)$ is an approximation of the gradient

$$\psi(t,\hat{\theta}(t-1)) = \left. \frac{d}{d\theta}\hat{y}(t \mid \theta) \right|_{\theta=\hat{\theta}(t-1)} \tag{20}$$

and $\varepsilon(t)$ in an approximation of

$$\varepsilon(t,\hat{\theta}(t-1)) = y(t) - \hat{y}(t \mid \hat{\theta}(t-1)) \tag{21}$$

Moreover, ℓ'_ε and $\ell''_{\varepsilon\varepsilon}$ are the derivatives of ℓ with respect to ε.

12.2. The special case of linear regressions

Let us now turn to the linear regression case (4). If we choose

$$\ell(\varepsilon, k) = \alpha_k \varepsilon^2 \tag{22}$$

in the criterion (16) it becomes a quadratic function of θ, and it is well known how it can be explicitly and exactly minimized by the *recursive least squares* (RLS) algorithms

$$\hat\theta(t) = \hat\theta(t-1) + \alpha_t R^{-1}(t)\varphi(t)\varepsilon(t) \tag{23}$$

$$R(t) = \lambda(t)R(t-1) + \alpha_t\varphi(t)\varphi^T(t) \tag{24}$$

$$\varepsilon(t) = y(t) - \varphi^T(t)\hat\theta(t-1) \tag{25}$$

In this we recognize the general algorithm (18) - (21) in our special case.

To bring out the connections to the "classical" low gain stochastic approximation (SA) algorithms let us introduce

$$\mu(t) = \left[\sum_{k=1}^{t} \beta(t, k)\right]^{-1} \tag{26}$$

Then

$$\bar R(t) \triangleq \mu(t)R(t) = \left[\sum_{k=1}^{t} \beta(t, k))\right]^{-1} \sum_{k=1}^{t} \beta(t, k)\varphi(k)\varphi^T(k)$$

becomes a normalized version of the Hessian of the criterion (16) (that is, it becomes an estimate of $E\varphi(t)\varphi^T(t)$). We can thus rewrite (23)-(24) as

$$\hat\theta(t) = \hat\theta(t-1) + \alpha_t\mu(t)\bar R^{-1}(t)\varphi(t)\varepsilon(t) \tag{27}$$

$$\bar R(t) = \bar R(t-1) + \mu(t)[\alpha_t\varphi(t)\varphi^T(t) - \bar R(t-1)] \tag{28}$$

The closer the $\lambda(j)$ are to 1, the smaller is $\mu(t)$. In particular, a constant "forgetting factor" $\lambda(j) \equiv \lambda < 1$ gives after a transient, a constant

$$\mu = (1 - \lambda) \tag{29}$$

Also, a constant $\lambda(j) \equiv 1$ gives the traditional "decreasing gain" algorithm

$$\mu(t) = \frac{1}{t} \tag{30}$$

All this means that the RLS algorithm (27)-(28) is brought into contact with the traditional SA-format, as treated in the earlier chapters of this book.

The matrix $\bar{R}^{-1}(t)$ in (27) modifies the update direction from the gradient direction to the Newton direction. A much-used alternative for tracking algorithms is the so called Least Mean Squares (LMS) algorithm, [13] which uses

$$\hat{\theta}(t) = \hat{\theta}(t-1) + \mu(t)\varphi(t)\varepsilon(t) \tag{31}$$

or, a normalized version (NLMS)

$$\hat{\theta}(t) = \hat{\theta}(t-1) + \frac{\mu(t)}{1 + \mu(t) \mid \varphi(t) \mid^2} \varphi(t)\varepsilon(t) \tag{32}$$

A fitting, 'systematic term' for this algorithm would be "stochastic gradient" (SG) algorithm.

12.3. A Bayesian approach - The Kalman Filter

The approach to deal with time varying systems via a weighted criterion (16) is - as we pointed out - *ad hoc*. A much more systematic approach would be to assume a model for how the changing parameters indeed vary. A simple such assumption would be that the parameters obey a *random walk*.

We thus have in the linear regression case

$$\theta_0(t) = \theta_0(t-1) + w(t) \tag{33}$$

$$y(t) = \theta_0^T(t)\varphi(t) + e(t). \tag{34}$$

We here assume $\{e(t)\}$ to be white Gaussian noise with variance $R_2(t)$, while $\{w(t)\}$ is white Gaussian noise with covariance matrix $R_1(t)$. We can then exactly compute the posterior density of the parameter $\theta_0(t)$ given the observations $y(s)$ and $\varphi(s)$ $s \leq t$. It is well known, see, e.g. Section 2.3 in [10], that the estimate $\hat{\theta}(t)$ that minimizes the conditional expectation, given past observations

$$\Pi(t) = E(\hat{\theta}(t) - \theta_0(t))(\hat{\theta}(t) - \theta_0(t))^T \tag{35}$$

(even in a matrix sense) is given by the Kalman filter, [6]

$$\hat{\theta}(t) = \hat{\theta}(t-1) + L(t)\varepsilon(t) \tag{36}$$

$$\varepsilon(t) = y(t) - \varphi^T(t)\hat{\theta}(t-1) \tag{37}$$

where the gain vector $L(t)$ is given by

$$L(t) = \frac{P(t-1)\varphi(t)}{\hat{R}_2(t) + \varphi^T(t)P(t-1)\varphi(t)} \tag{38}$$

and the matrix $P(t)$ is updated according to

$$P(t) = P(t-1) - \frac{P(t-1)\varphi(t)\varphi^T(t)P(t-1)}{\hat{R}_2(t) + \varphi^T(t)P(t-1)\varphi(t)} + \hat{R}_1(t),$$

$$P(0) = P_0, \tag{39}$$

We have here used the notations $\hat{R}_1(t)$ and $\hat{R}_2(t)$ to indicate that the values used in the algorithm may very well differ from the true values $R_1(t)$ and $R_2(t)$. In the case $\hat{R}_2(t) \equiv R(t)$ and $\hat{R}_2(t) \equiv R_2(t)$, however, $\hat{\theta}(t)$ is the conditional expectation of $\theta_0(t)$, given the observations $\{\varphi(k), y(k)\}$, $k \leq t$, and $P(t)$ is the conditional covariance matrix of the parameter estimation error. Moreover, the conditional distribution is Gaussian.

Note also that if $R_1(t)$ is known then (36)-(39) is the otpimal algorithm also for abrupt changes in the parameter θ_0. (Take $R_1(t) = 0$ except when a jump occurs, say for $t \in T_1$ take then $R_1(t) = R_1$.) However, this requires the time instants for the jumps to be known, not too realistic an assumption.

12.4. RLS and LMS in the light of the Bayesian approach

The Bayesian Approach (36)-(39) depends explicitly on the assumed incremental variance $\hat{R}_1(t)$ of the true parameters. It is not surprising that the success in the tracking by this algorithm depends on how well $\hat{R}_1(t)$ approximates the true variance $R_1(t)$.

It is then interesting to note that if we choose

$$
\begin{aligned}
\hat{R}_1(t) &= (\frac{1}{\lambda(t)} - 1) \times \left[P(t-1) - \frac{P(t-1)\varphi(t)\varphi^T(t)P(t-1)}{\lambda(t)/\alpha_t + \varphi^T(t)P(t-1)\varphi(t)} \right] \\
&\approx \left(\frac{1}{\lambda(t)} - 1 \right) P(t-1)
\end{aligned}
\tag{40}
$$

$$\hat{R}_2(t) = \lambda(t)/\alpha_t \tag{41}$$

then the Kalman filter (36)-(39) gives exactly the RLS algorithm (23)-(25). (To see this, we could first rewrite (24), using the matrix inversion lemma as $P(t) = R^{-1}(t)$

$$P(t) = \frac{1}{\lambda(t)}[P(t-1) - \frac{P(t-1)\varphi(t)\varphi^T(t)P(t-1)}{[\lambda(t)/\alpha_t] + \varphi^T(t)P(t-1)\varphi(t)}] \tag{42}$$

See, e.g. Chapter 2 of [10]).

We will also note that the following choice in (36)-(39)

$$\hat{R}_1(t) = \mu^2 \frac{\varphi(t)\varphi^T(t)}{1 + \mu \mid \varphi(t) \mid^2} \tag{43}$$

$$\hat{R}_2(t) = 1 \tag{44}$$

$$P(0) = \mu \cdot I \tag{45}$$

leads to the normalised LMS algorithm (32).

The fact that the two most common algorithms for tracking the parameters of a linear regression both are special cases of the "optimal" one and correspond to different assumptions about the variations in the true parameter has several important consequences. For one, it makes that the traditional question about which is "better" RLS or LMS is meaningless - it all depends on the actual parameter changes in the particular application at hand.

12.5. The general nonlinear regression case revisited

To put the general model (12) more in line with the linear regression case, treated in Sections 12.2 - 12.4, we can make an approximate derivation of a general algorithm as follows.

Consider the general structure (12) together with a random walk model for the variation of the "true parameter vector":

$$\begin{aligned} \theta_0(t) &= \theta_0(t-1) + w(t) \\ y(t) &= \hat{y}(t \mid \theta_0(t)) + e(t). \end{aligned} \tag{46}$$

Suppose that we have an approximation $\theta_*(t)$ of $\theta_0(t)$ available. We can then write, using the mean value theorem

$$\hat{y}(t \mid \theta_0(t)) = \hat{y}(t \mid \theta_*(t)) + (\theta_0(t) - \theta_*(t))^T \psi(t, \xi(t)) \tag{47}$$

where $\xi(t)$ is a value "between" $\theta_*(t)$ and $\theta_0(t)$. Here $\psi(t, \theta)$ is the gradient of $\hat{y}(t \mid \theta)$, as defined in (20). Normally, $\psi(t, \xi(t))$ would not be known, but we may assume that an approximation

$$\psi(t) \approx \psi(t, \xi(t)) \tag{48}$$

is available. Introduce the known variable

$$z(t) = y(t) - \hat{y}(t \mid \theta_*(t)) + \theta_*^T(t)\psi(t). \tag{49}$$

Subject to the approximation (48) we can then rewrite (46) as

$$\begin{aligned} \theta_0(t) &= \theta_0(t-1) + w(t) \\ z(t) &= \theta_0^T(t)\psi(t) + e(t) \end{aligned} \tag{50}$$

and we are back to the situation of Section 12.2. A natural choice $\theta_*(t)$ of a good approximation of $\theta_0(t)$ would be the previous estimate $\theta_*(t) = \hat{\theta}(t-1)$. We then obtain algorithms of the recursive prediction error type since

$$z(t) - \hat{\theta}^T(t-1)\psi(t) = y(t) - \hat{y}(t \mid \hat{\theta}(t-1)). \tag{51}$$

As $\hat{\theta}(t-1)$ comes closer to $\theta_*(t)$, the approximation involved in going from (46) to (50) will become arbitrarily good. This shows that an asymptotic theory of tracking parameters in arbitrary model structures can be developed from the linear regression case.

It should also be noted that in the non-linear regression case (12), it may be beneficial to let the gain matrix P in (38) be affected also by cross terms that reflect the uncertainty of the estimates of internal "states". If the prediction/parameter estimation problem inherent in (46) is described by an extended state vector (containing both the system's states and the vector θ) we obtain a description like

$$\begin{pmatrix} x(t+1) \\ \theta(t+1) \end{pmatrix} = \begin{pmatrix} A(\theta(t))x(t) \\ \theta(t) \end{pmatrix} + \begin{pmatrix} B(\theta(t))u(t) \\ 0 \end{pmatrix} \begin{pmatrix} v(t) \\ w(t+t) \end{pmatrix} \quad (52)$$

$$y(t) = C(\theta(t))x(t) + e(t) \quad (53)$$

The estimation of the extended state

$$X(t) = \begin{pmatrix} x(t) \\ \theta(t) \end{pmatrix} \quad (54)$$

can now be approached by non-linear filtering techniques, such as the extended Kalman filter. A careful analysis shows that the resulting algorithm for updating $\hat{\theta}(t)$ is of the recursive prediction error family (18) - (19), (provided the dependence of the "Kalman gain" on θ is properly accounted for). See [10] for such a discussion. However, the filtering approach gives a more complicated expression for $R^{-1}(t) = P(t)$ in (19) in that the cross covariance matrix for $\hat{x}$ and $\hat{\theta}$ also enters. While these terms have no asymptotic effect as the gain tends to zero, they may very well have positive transient effects. This still has to be carefully analysed.

12.6. Some other assumptions about the changing parameters

Global and local trends:

In some cases we may know that the parameter changes typically show trends, so that they continue for a while in a certain direction. To capture this we may model them as

$$\theta_0(t) = \theta_0(t-1) + \gamma v(t) + \delta(t) \quad (55)$$

where $\{v(t)\}$ is a correlated stochastic process and $\delta(t)$ is a deterministic or slowly varying stochastic vector. The term $\delta(t)$ models the global trends while $\{v(t)\}$ describes the local trends with the amount of correlation in $\{v(t)\}$ determining the duration of the local trends.

When $\delta(t)$ can be described as a random walk (possibly with zero increments) and $v(t)$ can be modelled as filtered white noise equation (55) can be rewritten as

$$X(t) = A(t)X(t-1) + w(t) \quad (56)$$
$$\theta_0(t) = CX(t) \quad (57)$$

$$Ew(t)w^T(s) = \begin{cases} R_1(t), & t = s \\ 0, & t \neq s \end{cases} \quad (58)$$

where

$$X(t) = \left(\begin{array}{c} \theta_0(t) \\ x(t) \end{array} \right). \tag{59}$$

Furthermore

$$A(t) = \left(\begin{array}{cc} I & D(t) \\ 0 & a(t) \end{array} \right), \quad R_1(t) = \left(\begin{array}{cc} 0 & 0 \\ 0 & \bar{R}_1(t) \end{array} \right) \tag{60}$$

where the matrix elements $D(t)$, $a(t)$ and $\bar{R}_1(t)$ come from the description of $v(t)$. Clearly (11) is a special case of (55)-(57). Combining this description with (4) gives

$$x(t) = A(t)X(t-1) + \gamma w(t) \tag{61}$$
$$y(t) = [\varphi^T(t) \ 0]X(t) + e(t). \tag{62}$$

This is still an estimation problem for which the Kalman filter gives the optimal solution (provided w and e are Gaussian with known covriances). One can immediately write down the filter and read the $\hat{\theta}(t)$-update formula from the upper part of the $\hat{X}(t)$ expression. This approach has been termed *multistep* algorithms by [7], and [11] and [12]. See also [2].

Abrupt changes

A typical situation may be that the dynamics remains constant for a while, and then suddenly goes through a change at a random time instant. To capture this we may describe $\theta_0(t)$ as

$$\theta_0(t) = \theta_0(t-1) + w(t) \tag{63}$$

$$w(t) = \left\{ \begin{array}{ll} 0 & \text{with probability } 1 - \gamma^2 \\ v & \text{with probability } \gamma^2 \end{array} \right. \tag{64}$$

where v is a random variable with some distribution. Furthermore, $w(t)$ and $w(s)$ are assumed to be independent for $t \neq s$. If v is zero mean with covariance matrix R_1, $w(t)$ will have the covariance matrix $\gamma^2 R_1$. This type of behaviour occurs for example in signal segmentation problems.

One possibility to deal with systems subject to abrupt changes is to use the formulation of Section 12.3 with a time varying R_1. The fundamental problem then is that we do not know the time instants T_1 when the jumps occur. Estimating $R_1(t)$ thus becomes a problem of estimating T_1, which really is a detection problem. Detecting the time instants when the system parameters jump has been discussed extensively by [1]. [4] has used carefully designed change detection algorithms to supply (39) with as correct $\hat{R}_1(t)$ matrices as possible, and [5] discusses how to estimate R_1 at the jumps.

Let us now turn to another way of dealing with abrupt system changes, that is not based on direct estimation of $R_1(t)$ (or T_1) in (39). Consider the formulation (63) for sudden changes in the parameters. If v is described as a

Gaussian random variable with zero mean and covariance R_1, we can describe $w(t)$ as a sequence of Gaussian random variables with covariances $R_1(t)$, where $R_1(t)$ is either 0 or R_1, but we do not know when. We do know, however that, for N data points, the true sequence $R_1(t)$ is one of 2^N possible combinations of 0 and R_1. In principle, we could run all the 2^N possible versions of (36)-(39), and we would know that the optimal $\hat{\theta}(t)$'s would be one of the obtained 2^N variants. How would we know which one? It is reasonable to assume that it would be the one that produced the smallest sum of squared prediction errors, $\varepsilon(t) = y(t) - \varphi^T(t)\hat{\theta}(t-1), t-1, \ldots, N$. That would at least be the maximum likelihood estimate among this finite collection of possibilities.

Let us introduce a slight reformulation of the problem (63) to the case where

$$\theta_0(t) = \begin{cases} \theta_0(t-1) & w.p. \ 1-\gamma^2 \\ v & w.p. \ \gamma^2 \end{cases} \tag{65}$$

This way of describing the abrupt change will be a quite acceptable alternative to (63) in most cases. [3] has shown that under (65) the ML estimate of the jump instants can be computed by examining only N (rather than 2^N) of the possible values. Further reductions to a constant number of branches can be obtained at the price of a certain risk of missing the global ML-estimate. However, a test is always possible to perform, that can tell that the obtained estimate indeed is the global ML one.

§13 Asymptotic properties of the decreasing gain case

The actual use of the adaptive algorithms is to track time-varying properties of a system or a signal and we shall devote most of this part to that problem. Still, a natural first question is to ask how well the algorithms are capable to handle a time invariant system. This corresponds to the special case $R_1(t) = \hat{R}_1(t) = 0$ in (39), or $\lambda(j) \equiv 1$ in (24).

A substantial part of [10] is devoted to such analysis, and we shall here only quote the bottom lines:

(a) A recursive prediction error algorithm (18) will, as t tends to infinity, and as the gain tends to zero converge to a local minimum of the expected loss function

$$\bar{V}(\theta) = E\ell(\varepsilon(t,\theta), t) \tag{66}$$

i.e.

$$\hat{\theta}(t) \to \arg\min \bar{V}(\theta) \qquad \text{w.p 1 as } t \to \infty \tag{67}$$

(b) If, in addition, the Gauss-Newton search direction (18)–(19) is used, and asymptotically equal weighting is used ($\lambda(j) \equiv 1$) then the asymptotic accuracy

$$\bar{P} = \lim_{t\to\infty} tE(\hat{\theta}(t) - \theta_0)(\hat{\theta}(t) - \theta_0)^T$$

will be the same as for the corresponding off-line estimation method.

These asymptotic properties are thus the best one could ask for. It remains though to study how the algorihtms actually can cope with time varying systems. This is the question we turn to next.

§ 14 Estimation of the tracking ability of the algorithms

In the analysis of the tracking ability we shall only study algorithms for linear regressions. We first develop an exact expression for the parameter error.

We write the basic tracking algorithm as

$$\hat{\theta}(t) = \hat{\theta}(t-1) + \mu P_t L(\varphi(t))(y(t) - \varphi^T(t)\hat{\theta}(t-1)) \tag{68}$$

It is straightforward to see how this relates to the different algorithms we have studied. Moreover, we describe the variations of the true parameter vector as

$$\theta_0(t) = \theta_0(t-1) + \gamma \cdot w(t) \tag{69}$$

where we will use γ as a scaling parameter. Let us introduce the parameter error

$$\tilde{\theta}(t) = \theta_0(t) - \hat{\theta}(t)$$

It is our objective to evaluate the size of this error as measured by the covariance matrix

$$\Pi(t) = E\tilde{\theta}(t)\tilde{\theta}^T(t) \tag{70}$$

where expectation is over all relevant random variables (see below).

14.1. The approximate expression for the error covariance

From (68)–(69) we immediately obtain

$$\tilde{\theta}(t) = (I - \mu P_t L(\varphi(t))\varphi^T(t))\tilde{\theta}(t-1) + \gamma w(t) - \mu P_t L(\varphi(t))e(t) \tag{71}$$

Introduce the traditional assumptions

(i)

$$\{w(t)\} \text{ is a sequence of independent random vectors}$$
$$\text{with zero mean and } Ew(t)w^T(t) = R_1 \tag{72}$$

(ii)

$$\{e(t)\} \text{ is a sequence of independent random variables}$$
$$\text{with zero mean and } Ee^2(t) = R_2 \tag{73}$$

(iii)

$$\{\varphi(t)\} \text{ is a sequence of random vectors with}$$
$$EL(\varphi(t))\varphi^T(t) = Q \tag{74}$$

Also, let

$$EL(\varphi(t))L^T(\varphi(t)) = Q_L \tag{75}$$

In fact, it is sufficient to let $\{w(t)\}$ and $\{e(t)\}$ be martingale differences. The matrices R_1, R_2 and Q could very well be time-varying with obvious changes for the sequel.

The conventional approximating expression for the covariance matrix is now

$$\hat{\Pi}(t) = (I - \mu P_t Q)\hat{\Pi}(t-1)(I - \mu P_t Q)^T + \gamma^2 R_1 + \mu^2 R_2 P_t Q_L P_t^T \tag{76}$$

Most often the analysis is given for $P_t = I$ and often also confined to the asymptotic case $t \to \infty$

$$\hat{\Pi}(t) \to \hat{\Pi}_*, \bar{P}_t \to P_*$$

where $\hat{\Pi}_*$ is the solution of

$$0 = -\mu P_* Q \hat{\Pi}_* - \mu \hat{\Pi}_* Q^T P_*^T + \mu^2 P_* Q_L \hat{\Pi}_* Q^T P^T + $$
$$\gamma^2 R_1 + \mu^2 R_2 P_* Q_L P_*^T \tag{77}$$

As an alternative the third term is omitted, to define the stationary covariance matrix $\hat{\Pi}_{**}$ as the solution of

$$P_* Q \hat{\Pi}_{**} + \hat{\Pi}_{**} Q^T P_*^T = \frac{\gamma^2}{\mu} R_1 + \mu R_2 P_* Q_L P_*^T \tag{78}$$

For the "recursive least squares case" $P_* = Q^{-1}$ we obtain for example

$$\hat{\Pi}_{**} = \frac{1}{2}(\frac{\gamma^2}{\mu} R_1 + \mu R_2 \cdot Q^{-1} Q_L Q^{-1}) \tag{79}$$

Expressions like these clearly bring out the choice of adaptation gain μ as a trade-off between tracking ability (large μ, first term) and noise sensitivity (small μ, second term).

The objective of the analysis is then to show that

$$\lim_{\mu \to 0} \lim_{t \to \infty} \Pi(t) = \hat{\Pi}_{**} \tag{80}$$

in some sense.

14.2. An exact expression for the error covariance

We now add to (72)-(74) the following assumption:

(iv)

$$P_t \quad \text{is deterministic and} \tag{81}$$
$$\{\varphi(t)\} \text{is a sequence of independent vectors,}$$
$$\text{mutually independent also of} \{e(t)\} \text{ and} \{w(t)\}. \tag{82}$$

Let us say right away that this assumption is violated for most applications of the model (68) to problems in control and signal processing. Anyway, it will bring out many useful features of the problem at hand.

The key result that (82) implies is that $P_t L(\varphi(t))\varphi^T(t)$ will become independent of $\tilde{\theta}(t-1)$. By multiplying each member of (71) by its transpose and taking expectation we thus obtain

$$\Pi(t) = \Pi(t-1) - \mu\bar{Q}\Pi(t-1) - \mu\Pi(t-1)\bar{Q}^T +$$
$$\mu^2 E P_t L(\varphi(t))\varphi^T(t)\tilde{\theta}(t-1)\tilde{\theta}^T(t-1)\varphi(t)L^T(\varphi(t))P_t$$
$$+\gamma^2 R_1 + \mu^2 R_2 \tilde{Q} \tag{83}$$

where

$$\bar{Q} = E P_t L(\varphi(t))\varphi^T(t) \tag{84}$$

$$\tilde{Q} = E P_t L(\varphi(t))L^T(\varphi(t))P_t \tag{85}$$

(In notation, we assume stationarity of $\varphi(t)$ and P_t. There would be no problem to use time-varying $\bar{Q}$ and $\tilde{Q}$ for most of the discussion.)

Note that (83) involves no approximation. To deal with the fourth term of the right hand side, we use Kronecker products $\otimes$ to write it as

$$\begin{aligned} \mathrm{col} \quad & \{E P_t L(\varphi(t))\varphi^T(t)\tilde{\theta}(t-1)\tilde{\theta}^T(t-1)\varphi(t)L^T(\varphi(t))P_t\} = \\ = \quad & E\{(I_d \otimes P_t L(\varphi(t))\varphi^T(t))(P_t L(\varphi(t))\varphi^T(t) \otimes I_d| \\ & \mathrm{col}\ (\tilde{\theta}(t-1)\tilde{\theta}^T(t-1)) = M \cdot \mathrm{col}\ \Pi(t-1) \end{aligned} \tag{86}$$

(The operation " col A" means that we form a column vector of the matrix A by stacking its columns on top of each other.

Here we define

$$M = E\{(I_d \otimes P_t L(\varphi(t))\varphi^T(t))(P_t L(\varphi(t))\varphi^T(t) \otimes I_d) \tag{87}$$

I_d is the identity matrix of size $d \times d$ and $d = \dim \theta$. let us also define

$$\tilde{M} = M - (I_d \otimes \bar{Q})(\bar{Q} \otimes I_d) \tag{88}$$

By applying "col" to (83) we now obtain

$$\begin{aligned} \mathrm{col}\ \Pi(t) \ &= \ (I - \mu(I_d \otimes \bar{Q} + \bar{Q} \otimes I_d) + \mu^2 M)\ \mathrm{col}\ \Pi(t-1) \\ &= \ \gamma^2\ \mathrm{col}\ R_1 + \mu^2 R_2\ \mathrm{col}\ \tilde{Q} \end{aligned} \tag{89}$$

Let us, for simplicity introduce

$$S(\mu) = I - \mu(I_d \otimes \bar{Q} + \bar{Q} \otimes I_d) + \mu^2 M. \tag{90}$$

Then we can write the solution of (89) explicitly as

$$\mathrm{col}\ \Pi(t) = S(\mu)^t\ \mathrm{col}\ \Pi(0) + \sum_{k=1}^{t} S(\mu)^{t-k}\{\gamma^2\ \mathrm{col}\ R_1 + \mu^2 R_2\ \mathrm{col}\ \tilde{Q}\} \tag{91}$$

Again, *this is the exact expression* for the exact covariance matrix $\Pi(t)$ under the assumptions (72), (73), (74), (81), (82). No approximations and no small step size μ here!

If the eigenvalues of $S(\mu)$ are all inside the unit circle (as they surely are for small enough μ if $\bar{Q}$ is positive definite), then the limit of (90) exists as $t \to \infty$ and we have

$$\Pi(t) \to \Pi_* \qquad \text{as } t \to \infty \tag{92}$$

where

$$\text{col } \Pi_* = (I - S(\mu))^{-1}\{\gamma^2 \text{ col } R_1 + \mu^2 R_2 \text{ col } \tilde{Q}\} \tag{93}$$

Using (91) we obtain

$$\text{col } \Pi_* = (I_d \otimes \bar{Q} + \bar{Q} \otimes I_d - \mu M)^{-1}\{\frac{\gamma^2}{\mu} \text{ col } R_1 + \mu R_2 \text{ col } \tilde{Q}\} \tag{94}$$

Let us define the matrix $\hat{\Pi}_{**}$ by

$$\bar{Q}\hat{\Pi}_{**} + \hat{\Pi}_{**}\bar{Q} = \frac{\gamma^2}{\mu}R_1 + \mu R_2 \tilde{Q} \tag{95}$$

or, equivalently

$$\text{col } \hat{\Pi}_{**} = (I_d \otimes \bar{Q} + \bar{Q} \cdot I_d)^{-1}\{\frac{\gamma^2}{\mu} \text{ col } R_1 + \mu R_2 \text{ col } \tilde{Q}\} \tag{96}$$

Then we have

$$\text{col } \Pi_* = (I_{d^2} - \mu Q_2^{-1} M)^{-1} \text{ col } \hat{\Pi}_{**} \tag{97}$$

where we used, for short

$$Q_2 = I_d \otimes \bar{Q} + \bar{Q} \otimes I_d. \tag{98}$$

Once more, we stress that so far no approximations are involved. All expression hold for all μ, γ ((92) and onwards assume just $S(\mu)$ to be stable).

14.3. Relations between the exact and approximate expressions

We realize clearly that $\hat{\Pi}_{**}$ defined by (95) or (96) is the same as the traditional approximation (78), given that P_t is deterministic. The expression (97) is therefore an excellent basis for comparing this approximation with the exact expression. We have

$$| \text{ col } \Pi_* - \text{ col } \hat{\Pi}_{**}| \leq ||I_{d^2} - \mu(Q_2^{-1} M)^{-1}|| \text{ col } \hat{\Pi}_{**}| \tag{99}$$

Now, as soon as

$$\mu < ||Q_2^{-1} M||^{-1} \overset{\triangle}{=} \mu_0 \tag{100}$$

we can use the convergent expression

$$(I_{d^2} - \mu Q_2^{-1} M)^{-1} = I_{d^2} + \mu Q_2^{-1} M - \mu^2 Q_2^{-1} M Q_2^{-1} M + \mu^3 \cdots \tag{101}$$

and

$$\|I_{d^2} - (I_{d^2} - \mu Q_2^{-1} M)^{-1}\| \le \mu \cdot C_\mu \qquad (102)$$

with

$$C_\mu = \frac{1}{1 - \mu/\mu_0} \qquad (103)$$

so that

$$|\operatorname{col} \Pi_* - \operatorname{col} \hat{\Pi}_{**}| \le \mu \cdot C_\mu \cdot |\operatorname{col} \hat{\Pi}_{**}| \qquad (104)$$

The relative accuracy in the approximation $\hat{\Pi}_{**}$ thus improves with the factor μ as soon as $\mu < \mu_0$.

Note that (104) is an expression of "the kind" (80). However, it is considerably more explicit. It is really not asymptotic in $\mu \to 0$, since it is applicable for all $\mu < \mu_0$. It also shows that for small μ $\hat{\Pi}_{**}$ is a reasonable approximation of Π_* regardless of the relationship between μ and γ.

Based on (97) and (101) we can also provide a series expansion for Π_* in powers of μ:

$$\operatorname{col} \Pi_* = Q_2^{-1}(\frac{\gamma^2}{\mu} \operatorname{col} R_1 + \mu R_2 \operatorname{col} \tilde{Q})$$

$$+\mu Q_2^{-1} M Q_2^{-1} M Q_2^{-1}(\frac{\gamma^2}{\mu} \operatorname{col} R_1 + R_2 \operatorname{col} \tilde{Q})$$

$$-\mu^2 Q_2^{-1} M Q_2^{-1} \frac{\gamma^2}{\mu} \operatorname{col} R_1 + \mu R_2 \operatorname{col} \tilde{Q}) + \cdots$$

$$= \frac{1}{\mu}[\gamma^2 Q_2^{-1} \operatorname{col} R_1] + [\gamma^2 Q_2^{-1} M Q_2^{-1} \operatorname{col} R_1] +$$

$$+\mu[R_2 Q_2^{-1} \operatorname{col} \tilde{Q} - \gamma^2 Q_2^{-1} M Q_2^{-1} M Q_2^{-1} \operatorname{col} R_1] + \ldots \qquad (105)$$

In the formal expansion of Π_* in μ^k there are thus terms that are proportional to μ^0. This, however, does not conflict with the fact (104) that $\hat{\Pi}_{**}$ (which contains no such term) always approximates Π_* arbitrarily well for small enough μ.

Let us now turn to the transient expressions (86) and (91).

Assuming $P_t = P$ to be deterministic we have from (76)

$$\operatorname{col} \hat{\Pi}(t) = \hat{S}(\mu)^T \operatorname{col} \hat{\Pi}(0) + \sum_{k=1}^{t} \hat{S}(\mu)^{t-k}$$

$$(\gamma^2 \operatorname{col} R_1 + \mu^2 R_2 \operatorname{col} \tilde{Q}) = \hat{S}(\mu)^t \operatorname{col} \hat{\Pi}(0) +$$

$$(\hat{S}(\mu) - I)^{-1}(\hat{S}(\mu)^t - I)(\gamma^2 \operatorname{col} R_1 + \mu^2 R_2 \operatorname{col} \tilde{Q}) \qquad (106)$$

where

$$\hat{S}(\mu) = I - \mu(I_d \otimes \bar{Q} + \bar{Q} \otimes I_d) + \mu^2(I_d \otimes \bar{Q})(\bar{Q} \otimes I_d) \qquad (107)$$

According to (91) the exact expression for $\Pi(t)$ is the same with $S(\mu)$ replacing $\hat{S}(\mu)$. But, according to (90), (88) and (107)

$$S(\mu) - \hat{S}(\mu) = \mu^2 \tilde{M} \tag{108}$$

Just as in the stationary case, straightforward calculations show that (108) implies that

$$| \text{col } \Pi(t) - \text{col } \hat{\Pi}(t)| \leq 2 \cdot \mu \cdot | \text{col } \hat{\Pi}(t)| + C\mu(1-\mu)^t|\Pi(0)| \tag{109}$$

for $\mu \leq \mu_0$, (assuming that $\hat{\Pi}(0) = \Pi(0)$). The approximation is thus valid also in the transient phase.

14.4. An alternative approach

Practically the same result as (109) can be shown with a slightly different technique. We prove the following lemma:

Lemma 1. *Let $\Pi(t)$ be defined by*

$$\Pi(t) = A_\mu \Pi(t-1) A_\mu^T + \mu x + \rho(t), \quad \Pi(0) = \Pi_0 \tag{110}$$

where A_μ is a stable matrix:

$$\|A_\mu^t\| \leq C_A (1 - \alpha\mu)^{t/2} \tag{111}$$

and

$$|\rho(t)| \leq \mu \cdot \sigma(\mu)(|x| + c_\Pi \max_{t-\tau \leq k \leq t} |\Pi(k)|) \tag{112}$$

for some decreasing function $\sigma(\mu)$. Let $\hat{\Pi}(t)$ be defined as in (110) but without the term $\rho(t)$. Let μ_0 be such that $\sigma(\mu_0) \leq \frac{1}{2} C_A^2 C_\Pi/\alpha$. Then for $\mu \neq \mu_0$ we have

$$|\Pi(t) - \hat{\Pi}(t)| \leq C^* \sigma(\mu)|x| + C_* \sigma(\mu)(\sigma(\mu) + \mu t(1-\alpha\mu)^{t-\tau})|\Pi_0| \tag{113}$$

where

$$C^* = \frac{C_A^2}{\alpha}(1 + 4C_\Pi \frac{C_A^2}{\alpha}) \tag{114}$$

$$C_* = C_A^4 C_\Pi (1 + 2(C_A/\alpha)^2 \cdot C_\Pi) \tag{115}$$

Remark. Note that the "size" of $\hat{\Pi}(t)$ is

$$|\hat{\Pi}(t)| \sim |x| + (1 - \alpha\mu)^t|\Pi_0| \tag{116}$$

so (113) tells us that the relative approximation of $\Pi(t)$ by $\hat{\Pi}(t)$ improves with the factor $\sigma(\mu)$ as μ decreases.

Proof. Let

$$\tilde{\Pi}(t) = \Pi(t) - \hat{\Pi}(t) \tag{117}$$

$$\Pi_s(t) = \Pi(t) - A_\mu^t \Pi_0 (A_\mu^T)^t \tag{118}$$

$$\hat{\Pi}_s(t) = \hat{\Pi}(t) - A_\mu^t \Pi_0 (A_\mu^T)^t \tag{119}$$

Also define

$$m(t) = \max_{t-\tau \le k \le t} |\Pi(k)| \tag{120}$$

$$\tilde{m}(t) = \max_{k \le t} |\Pi_s(k)| \tag{121}$$

Then

$$m(t) \le \tilde{m}(t) + C_A^2 (1 - \mu\alpha)^{t-\tau} |\Pi_0| \tag{122}$$

$$\tilde{\Pi}(t) = \Pi_s(t) - \hat{\Pi}_s(t) \tag{123}$$

Thus

$$\tilde{m}(t) \le \max_{k \le t} |\hat{\Pi}_s(k)| + \max_{k \le t} |\tilde{\Pi}(k)| \tag{124}$$

Now

$$
\begin{aligned}
|\tilde{\Pi}(t)| &= |\sum_{k=1}^{t} A_\mu^{t-k} \rho(k)(A_\mu^T)^{t-k}| \le \\
&\le \sum_{k=1}^{t} C_A^2 (1 - \alpha\mu)^{t-k} \sigma(\mu)(C_x \cdot \mu \cdot |x| + C_\Pi \cdot \mu \cdot \tilde{m}(k) + \\
&+ C_\Pi \mu \cdot C_A^2 (1 - \alpha\mu)^{k-\tau} |\Pi_0|)
\end{aligned}
\tag{125}
$$

using (112) and (122). Thus

$$
\begin{aligned}
|\tilde{\Pi}(t)| &\le \frac{C_A^2}{\alpha} \sigma(\mu)(C_x|x| + C_\Pi \cdot \tilde{m}(t)) \\
&+ t\sigma(\mu)\mu C_\Pi C_A^4 (1 - \alpha\mu)^{t-\tau} |\Pi_0|
\end{aligned}
\tag{126}
$$

Moreover

$$|\hat{\Pi}_s(t)| = |\sum_{k=1}^{t} A_\mu^{t-k} \mu x (A_\mu^T)^{t-k}| \le \frac{C_A^2}{\alpha} \cdot |x| \tag{127}$$

We also have that

$$\max_{k \le t} k(1 - \alpha\mu)^{k-\tau} \le \frac{1}{\alpha\mu} \cdot (1 - \alpha\mu)^{\frac{1}{\alpha\mu} - \tau} \le \frac{1}{\alpha\mu}$$

(assuming $\tau < 1/\alpha\mu$). Inserting (126) and (127) into (124) gives

$$
\begin{aligned}
\tilde{m}(t) \le \frac{C_A^2}{\alpha} |x|(1 + \sigma(\mu)) + \frac{C_A^2}{\alpha} \cdot C_\Pi \sigma(\mu) \cdot \tilde{m}(t) + \\
\frac{1}{\alpha} \sigma(\mu) C_\Pi C_A^4 |\Pi_0|
\end{aligned}
\tag{128}
$$

Let μ_0 be defined by

$$\sigma(\mu_0) = \frac{1}{2}\frac{C_A^2}{\alpha} \cdot C_\Pi \tag{129}$$

Then, for $\mu \leq \mu_0$

$$\tilde{m}(t) \leq 2\frac{C_A^2}{\alpha}|x|(1 + \sigma(\mu_0)) + 2\sigma(\mu)C_\Pi C_A^4|\Pi_0|/\alpha \tag{130}$$

Inserting this into (126) now gives the desired result.   □

14.5. Relaxing the assumption of independent regressors

If we relax the assumption (82) (while retaining (72) – (74)) the calculations leading to the exact expression (89) are still valid except that there are three additional terms in the right hand side:

$$\alpha(t) = \mu E P_t L(\varphi(t))\varphi^T(t)\tilde{\theta}(t-1)\tilde{\theta}^T(t-1) - \mu\bar{Q}\Pi(t-1) \tag{131}$$
$$\alpha^T(t)$$

and

$$\tilde{\alpha}(t) = \text{col } \mu^2(E P_t L(\varphi(t))\varphi^T(t)\tilde{\theta}(t-1)\tilde{\theta}^T(t-1)\varphi(t)L^T(\varphi(t))P_t - M)$$

Now, $\varphi(t)$ and $\tilde{\theta}(t-1)$ are no longer independent so $\alpha(t)$ is not zero. However, if the dependence among $\{\varphi(t)\}$ decays sufficiently fast it seems reasonable that $\alpha(t)$ is of an order of magnitude less than $\mu \cdot \Pi(t-1)$. In that case we could still apply Lemma 5.1. We demonstrate the idea for the following special case:

(v) Assume that $\varphi(t)$ and $\varphi(s)$ are independent for $|t-s| \geq \tau$. Also assume that

$$|P_t L(\varphi(t))\varphi^T(t)| \leq C_\varphi \tag{132}$$

and that $\{\varphi(t)\}$ are independent of $\{e(t)\}$ and $\{w(t)\}$ in (10) – (11).

We shall take $\tau = 2$ in the calculations below. We have

$$\alpha(t) = \mu E P_t L(\varphi(t))\varphi^T(t)((I - \mu P_{t-1}L(\varphi(t-1)\varphi^T(t-1))\tilde{\theta}(t-2)$$
$$+\mu P_{t-1}L(\varphi(t-1))e(t-1) + \gamma w(t-1)(\ldots)^T =$$
$$= \mu\bar{Q}(\Pi(t-2) - \Pi(t-1)) + f_1(t) + f_2(t)$$

where f_1 takes care of the squared terms with $e^2(t-1)$ and $w^2(t-1)$ and where f_2 takes care of the cross terms with $\tilde{\theta}(t-1)$. We then have

$$|f_1(t)| \leq \mu^3 \cdot C_1 + \mu\gamma^2 \cdot C_2$$

$$|f_2(t)| \leq C_3\mu^2|\Pi(t-2)|$$

using in the last step (132) and

$$\mu E P_t L(\varphi(t))\varphi^T(t)\mu P_{t-1}L(\varphi(t-1))\varphi^T(t-1)\tilde{\theta}(t-2)\tilde{\theta}^T(t-2)|$$

$$\leq \mu^2 C_\varphi^2 E|\tilde{\theta}(t-2)\tilde{\theta}^T(t-2)|$$

and similarly for the other cross terms.

We also easily obtain from (83)

$$|\mu\bar{Q}(\Pi(t-2) - \Pi(t-1))| \leq C_5\mu^2|\Pi(t-2)| + \mu\gamma^2|R_1| +$$
$$\mu^3 R_2|\tilde{Q}| + \mu|\alpha(t)|$$

all which yields

$$|\alpha(t)| + |\tilde{\alpha}(t)| \leq C_6\mu^2|\Pi(t-2)| + C_7\mu^3 + C_8\mu\gamma^2 \qquad (133)$$

Lemma 1 now again gives the result (109).

One can imagine how to extend this to other finite τ. In [9] the result is extended to general mixing properties of $\{\varphi(t)\}$. The function $\sigma(\mu)$ in Lemma 1 will then depend on the mixing rate.

References for Part III

[1] M. Basseville and A. Benveniste. *Detection of Abrupt Changes in Signals and Dynamical Systems*. Lecture Notes in Control and Information Sciences. Springer–Verlag, 1986

[2] A. Benveniste. Design of adaptive algorithms for the tracking of time–varying systems. *International Journal of Adaptive control and Signal Processing*, 1:3–29, 1987

[3] F. Gustafsson. Optimal segmentation of linear regression parameters. Technical report, Department of Electrical Engineering, Linköping University, 1990.

[4] T. Hägglund. Adaptive control of systems subject to large paremeter changes. In *Proc. 9th IFAC World Congress*, 1984.

[5] J. Holst and N. K. Poulsen. Self tuning control of plants with abrupt changes. In *Proc. 9th IFAC World Congress*, 1984.

[6] R. E. Kalman and R. S. Bucy. New results in linear filtering and prediction theory. *Transaction ASME, Journal of Basic Engineering*, 1961.

[7] A. P. Korostelev. Multistep procedures of stochastic optimization. *Avtomatikha i Telemekhanika*, (5):82–90, 1981.

[8] L. Ljung. *System Identification - Theory for the User*. Prentice-Hall, Englewood Cliffs, N. J., 1987.

[9] L. Ljung and P. Priouret. A result of the mean square error obtained using general tracking algorithms. *Int. J. of Adaptive Control*, 5:231–250, 1991.

[10] L. Ljung and T. Söderström. *Theory and Practice of Recursive Identification*. MIT press, Cambridge, Mass., 1983.

[11] S. V. Shilman and A. I. Yastrebov. Convergence of a class of multistep stochastic adaptation algorithms. *Avtomatikha i Telemekhanika*, 1976.

[12] S. V. Shilman and A. I. Yastrebov. Properties of a class of multistep gradient and pseudogradient algorithms of adaptation and learning. *Avtomatikha i Telemekhanika*, 1978.

[13] B. Widrow and M. E. Hoff Jr. Adaptive switching circuits. *IRE WESCON Convention Record*, 1960.

Previously published in the series DMV Seminar:

Volume 1: **Manfred Knebusch/Winfried Scharlau, Algebraic Theory of Quadratic Forms.**
1980, 48 pages, softcover, ISBN 3-7643-1206-8.

Volume 2: **Klas Diederich/IngoLieb, Konvexitaet in der Komplexen Analysis.**
1980, 150 pages, softcover, ISBN 3-7643-1207-6.

Volume 3: **S. Kobayashi/H. Wu/C. Horst, Complex Differential Geometry.**
2nd edition 1987, 160 pages, softcover, ISBN 3-7643-1494-X.

Volume 4: **R. Lazarsfeld/A. van de Ven, Topics in the Geometry of Projective Space.**
1984, 52 pages, softcover, ISBN 3-7643-1660-8.

Volume 5: **Wolfgang Schmidt, Analytische Methoden für Diophantische Gleichungen.**
1984, 132 pages, softcover, ISBN 3-7643-1661-6.

Volume 6: **A. Delgado/D. Goldschmidt/B. Stellmacher, Groups and Graphs: New Results and Methods.**
1985, 244 pages, softcover, ISBN 3-7643-1736-1.

Volume 7: **R. Hardt/L. Simon, Seminar on Geometric Measure Theory.**
1986, 118 pages, softcover, ISBN 3-7643-1815-5.

Volume 8: **Yum-Tong Siu, Lectures on Hermitian-Einstein Metrics for Stable Bundles and Kaehler-Einstein Metrics.**
1987, 172 pages, softcover, ISBN 3-7643-1931-3.

Volume 9: **Peter Gaenssler/Winfried Stute, Seminar on Empirical Processes.**
1987, 114 pages, softcover, ISBN 3-7643-1921-6.

Volume 10: **Jürgen Jost, Nonlinear Methods in Riemannian and Kaehlerian Geometry.**
2nd edition1991, 154 pages, softcover, ISBN 3-7643-2685-9.

Volume 11: **Tammo tom Dieck/Ian Hambleton, Surgery Theory and Geometry of Representations.**
1988, 122 pages, softcover, ISBN 3-7643-2204-7.